# COMPOST AFTER READING

# COMPOST AFTER READING

A PRACTICAL MANIFESTO FOR PURPOSEFUL DECOMPOSITION

CASSANDRA MARKETOS

ILLUSTRATIONS BY
SLUDGE THUNDER

Timber Press ◊ Portland, Oregon

Photos courtesy of Cassandra Marketos.
Illustrations by Sludge Thunder.



Timber Press
Workman Publishing
Hachette Book Group, Inc.
1290 Avenue of the Americas
New York, New York 10104
timberpress.com

Timber Press is an imprint of Workman Publishing, a division of Hachette Book Group, Inc. The Timber Press name and logo are registered trademarks of Hachette Book Group, Inc.

Printed in Illinois (VER) on responsibly sourced paper
Text and cover design by Hillary Caudle
Cover art by Sludge Thunder



ISBN 978-1-64326-462-2

A catalog record for this book is available from the Library of Congress.

# CONTENTS

PART ONE

# BEGINNINGS

# INTRODUCTION

I learned to compost when I was a child. Kind of. What I mean is that I learned how to add food scraps to the heap after dinner. I knew that my mom stood over the pile with a pitchfork sometimes, possibly turning it, and that a lizard lived somewhere in the depths that we named Fred. I never knew anything *specific* about how compost worked. It was just the way we handled our waste.

When I was much older, I volunteered at a plant nursery in Los Angeles, where one of my colleagues had a shabby bin out back that she used for experimenting with compost. I remember asking her questions: How does this work? What do you do to make it compost and not just straw and apple cores? What *is* compost, exactly? The process felt both specialized and highly mysterious. It's hard to describe the total disconnect that I felt watching her turn the pile in front of me, as though this were something that I was not allowed into.

Early attempts to demystify things were not successful. Ratios, equations, and tables of different materials and their chemical compositions were not the right entry point for me. I tried asking questions of friends, but my naivete was often frustrating. I also got conflicting information. One person said I could add citrus; another said definitely not. Bread was a questionable ingredient for some and a given for others. Some people turned their compost every few days, while others said to leave it alone for long stretches. How could so many things be true at once?

Eventually, I gave up and just started piling food scraps in the corner of my yard. I buried the scraps in leaf litter, wet it all with a hose, and left it there. I occasionally added fresh food scraps. Following the scent of some vague intuition, I would dig a small hole in the center of the pile, place the scraps delicately inside, and bury them again in fresh leaf litter. Sometimes I just covered the pile in dirt. It rained often that winter, keeping the pile consistently moist.

After some time passed, I began to notice things. If my pile lacked a certain amount of leaves or twigs, the food scraps would not decay as quickly. Instead, they would clump together, become slimy, and start to smell. I started adding more woody materials in ratio to my food scraps, maintaining an intuitive ratio. Another thing I noticed was that how dry the pile was seemed to impact how quickly stuff broke down. If it didn't rain enough, things slowed to a crawl. Zucchini ends would linger for weeks. I started splashing the pile with a hose. Shortly after that, I noticed that my compost was shrinking. I was adding voluminous amounts of food scraps, cardboard, and yard waste, but the pile was always staying the same size—if not getting smaller, at times. It was all decomposing so quickly. It was also beginning to smell. Not a bad smell, but a good one. When I opened the top of the heap up to add new food scraps, I was met with an aroma like deep, rich earth. Worms were starting to appear. Not just within the pile itself, but all across my tiny garden. My compost was working.

Compost made sense once I had my own pile on hand to interact with. I needed to be able to look and see what was going on, but also to tend it over time. That relationship opened the door for other things to start making sense. These days, I know much more about how compost works. I understand what ratio of nitrogen to carbon will create efficient decomposition and which bacteria are at work and how they break stuff down. I know that a well-tended pile will not stink or attract animals. I know that you can, in fact, compost citrus and bread, but there are reasons some people don't. It all just depends. Most important, though, I know that composting does not have to be difficult or even particularly complex. Anybody can start doing it, at any time, armed with just your eyes, your nose, and a little bit of common sense.

I wrote this book because I want to share this learning experience with others who have struggled to find their way into making compost. I want to give you a way to get started that isn't about memorizing formulas or a table of percentages, but about building a relationship to your own unique pile. I want to show you what you can learn by looking and touching and growing to understand something over time.

What can we learn by simply noticing?

I love my compost now. It is filled with bugs and beetles, worms, and even tiny salamanders. Under a microscope, it is positively luminous. Dozens of tiny creatures—some round and some tubular, some like corkscrews and some like waving fronds—zip about in their enviable work of turning "waste" into living, fertile soil. Just patting a handful of my compost around the

base of a plant results in noticeably improved vigor: the stalks become more taut, and leaves begin to spread with fresh and magnificent force.

Compost is everything. It's fun, and it's gross. It's scientific but also outright poetic. It's iterative and ritualistic, intimate and interconnected. It involves caring (one "tends" to their compost) but also brute strength. I've never been stronger than while I have been composting. Very few things play to this specific mixture of human needs—to be forceful but also loving—but compost is one of them. It feels like an inherently compassionate practice, helping one seek and inhabit that luminous space where life and death meet each other.

# A BRIEF HISTORY OF COMPOST

Humans have been composting for as long as we've existed. We have discovered and rediscovered the value of our waste across the centuries, often through methods as simple as observing death. While browsing *Compost Magazine*, I discovered this quote dating back to the beginning of the second century, by the Greek philosopher Plutarch:

> *Nevertheless, it is said that the people of Massalia fenced their vineyards round with the bones of the fallen, and that the soil, after the bodies had wasted away in it and the rains had fallen all winter upon it, grew so rich and became so full to its depths of the putrefied matter that sank into it, that it produced an exceeding great harvest in after years, and confirming the saying of Archilochus that "fields are fattened" by such a process.*

All the ingredients for a vigorous compost are present: the organic body, the leafy waste of the vines, the rain, and the ground. One wonders what Archilochus must have witnessed a thousand years prior—which fields were fattened and by whose corpses. These details of decomposition are lost to recorded history.

A set of clay tablets recovered from the time of the Akkadian Empire, thousands of years before Plutarch's time, document what some believe to be a recipe for making compost. Similar references have been found in ancient Hindu texts. In China, a composting method involving oil cake amended with crop residue was recorded by Chen Fu around the year 1149 BCE. Later, around 385 BCE, Greek warrior and philosopher Xenophon advised farmers to gather weeds and allow them to rot in water, in order to create a green manure for the improvement of their fields.

Early Hebrew scripture references manure being mixed with street sweepings and organic refuse, forming a dung hill that was kept at the edge of the city, like an early version of a town dump. The Talmud describes the process: "They lay dung to moisten and enrich the soil; dig about the roots of trees; pluck up the seckers; take off the leaves; sprinkle ashes; and smoke under the trees to kill vermin." Cleopatra, in 50 BCE, is reported to have made worms sacred after observing their connection to the cultivation of healthy soils. She enacted laws to make their removal from Egypt a crime punishable by death.

George Washington is referred to as "America's first composter." His preoccupation with finding ways to improve the poor soils at his home in Virginia led him to build a dung repository near his stables, where he brewed a mixture of manure and plant material to use as fertilizer. You can still tour a reconstruction of this site today. However, there is ample evidence that people were composting in America long before Washington. According to historian Jim Loewen, "Composting first appears in the historical record of what is now the United States in 1621, when Squanto showed the 'Pilgrims' how to put a fish in each corn hill, so the maize and squash would thrive." He goes on to describe how different pieces of Indigenous languages outlined the use and meaning of compost for Native peoples: "Narragansetts called the fish 'munnawhatteaûgs,' which means 'fertilizer' or 'that which enriches the land,' a word the English corrupted into 'menhaden.' The Abenakis of

Maine called them 'pauhagens,' which also means 'fertilizer,' a name the English shortened to 'pogies.'"

The history of compost is woven in this way through centuries. Every culture, at every turn, has arrived at the conclusion that dead and decaying matter is a vital source of life. It appears in our written records, our language, and, on occasion, the stone beneath our feet. For example, fossil evidence survives that suggests that ancient Scots planted their crops directly into composts made from manure, beginning about twelve thousand years ago.

John Adams was enthusiastic about manure and documented his preoccupation with it quite thoroughly in his diaries, which have been made available through the Massachusetts Historical Society.

> *In one of my common Walks, along the Edgeware Road, there are fine Meadows, or Squares of grass Land belonging to a noted Cow keeper. These Plotts are plentifully manured. There are on the Side of the Way, several heaps of Manure, an hundred Loads perhaps in each heap. I have carefully examined them and find them composed of Straw, and dung from the Stables and Streets of London, mud, Clay, or Marl, dug out of the Ditch, along the Hedge, and Turf, Sward cutt up, with Spades, hoes, and shovels in the Road. . . . This may be good manure, but it is not equal to mine.*

Compost also appears as a kind of myth, or even a dream. One summer, my friend David wrote to me from a trip. He had just heard something exciting about "witches" in medieval France who gathered nightly around compost. Apparently, they were using the piles for warmth on cool evenings. "*Los evangiles de ecreignes*, I think, but I'm busy and will need to look later." I translated the phrase and recorded it into a diary, but neither David nor I spoke of the witches again. We both forgot, distracted by other projects. It wasn't until months later that I remembered and returned to those pages, searching for more information. But my notes did not contain the translation that I so clearly recalled having included before. Instead, under the header "Compost witches," it read only: "Tea women of the people, who never yielded."

# WHAT IS COMPOST?

Every living thing on Earth will die.

That includes plants, animals, and humans. When we die, our bodies will begin to decompose. This process breaks large forms into smaller pieces and splits complex chemical arrangements into their individual molecules.

These components cycle back into the soil, forming the basis for all other growth. Roots grow. Shoots sprout. Leaves spread. Life, in all its infinite and complex capacities, regenerates. This is the fate of all organic matter.

It is also the operating principle of any compost pile. Composting is the breaking down of once-living organic matter into nutrient-rich, biologically active fertilizer.

There is, however, an important distinction between "compost" and the plain fact of decomposition: the presence of the composter. The composter is tasked with stewarding the conditions of the compost pile. They determine whether decomposition will be hastened or proceed aimlessly, whether it will take place in an enclosed chamber or open heap, and whether it will be driven primarily by bacteria or perhaps fungi. Thus, composting happens anywhere that a composter takes it upon themselves to deliberately manage the process of decomposition. This includes a great variety of forms and approaches—all, in my view, legitimized by their intention and none excluded by the range of their outcomes.

Composting is simple, but for the beginner it can feel distantly complex, even magical. I encourage you to embrace that feeling. The composter performs a kind of spell, I like to believe, casting deep into the world for something far greater than themselves.

# WHY COMPOST WORKS

Compost is a great gift to the soil. This is in large part due to something we call "humus." Humus is a dark, rich organic matter that is nutrient-dense, biodiverse, and smells richly of earth. It is the end product of decomposition; in other words, it is what remains when dead things can be broken down no further. It may seem like an ecological afterthought, mere leftovers, but humus is the whole point of your composting work. More than that, it could even be said that humus is the whole point of life on Earth. It is where we start from, in addition to where we will end.

It is humus that marks the difference between plain dirt and living soil, the thing that allows for the growth and flourishing of plants. It contains and releases elements and minerals, making them available for reuse as needed and in the precise quantities required for plants to thrive. Its microbial diversity works to break up toxins and suppress pathogens, and its molecular structure improves soil tilth, helping porous soils retain water and clay-like soils drain better. The range of its ability is matched only by the broad scope of its adaptability. Whatever is growing, whatever the conditions, humus will adjust to provide.

Exactly how humus does this is not totally understood. It cannot be identified by any singular molecular structure. Its innermost processes remain secret, having resisted scientific comprehension for centuries. In fact, the only thing consistent about humus is that it always seems to be different—to the point that some scientists wonder how to categorize it at all. "Humus cannot be regarded as a real substance," writes German soil scientist Erhard Hennig, "but rather as a process, a formation, built from a multitude of constantly changing factors." It can feel surreal that, when composting, we are casually creating such a remarkable, essential, and ultimately unknowable thing.

Humus occurs anywhere that organic material is able to accrue and decay, such as on forest floors, although it forms very slowly—in some cases, it can take a thousand years to form just 1 inch of humus. In places where it runs deepest, it forms a layer that is about a foot deep. Your backyard garden, if you have one, is not likely to contain an abundance of humus. The soil may be undernourished from decades of prior use or excessive use of synthetic fertilizers, or it may never have had much in the first place. That's where compost can come in handy. Good compost replenishes the soil, restoring its capacity to both contain and provide.

# HOW DECOMPOSITION WORKS

Broadly speaking, decomposition is a process of radical simplification. The complex molecules that combine to create all living matter are gradually broken apart into their individual components, in various and overlapping stages, through the action of decomposers, primarily microbes. These microbes exist in a variety of sizes and types. They feed directly on organic matter in some cases, or they can even feed on each other. Some of them secrete enzymes. Some of them secrete poop. Altogether, though, they have the highly visible effect of slowly degrading a once-living substance until it eventually transforms into the dark, earthlike matter that we call humus.

It follows that beginning composters often visualize decomposition as a form of absence, the steady and spontaneous decline of *something* into nothing. In fact, the opposite is true. Every aspect of decomposition is fueled by what something very much *alive* needs in order to survive and reproduce. When we talk about how to build "good" compost, then, what we are really discussing is how to build the ideal home for these decomposers.

Thus, it helps to have a sense of who exactly is at work in your pile. What do they eat, and how? What do they need to flourish? Knowing those details will help you hone your intuition about how to manage your pile.

## Aerobic bacteria

Aerobic bacteria are the first to show up in a compost pile, and they will multiply quickly in ideal conditions. To survive, they need carbon for energy, nitrogen to process that carbon, oxygen, and water. They help kick-start the breakdown process, producing an incredible variety of digestive enzymes that can break apart complex organic compounds like cellulose into simpler molecules. In doing so, they feed themselves but also release nutrients that become available for other microbes and plants.

Aerobic bacteria exist in your pile in huge numbers and with great variety. Under a microscope, some look like spheres, while

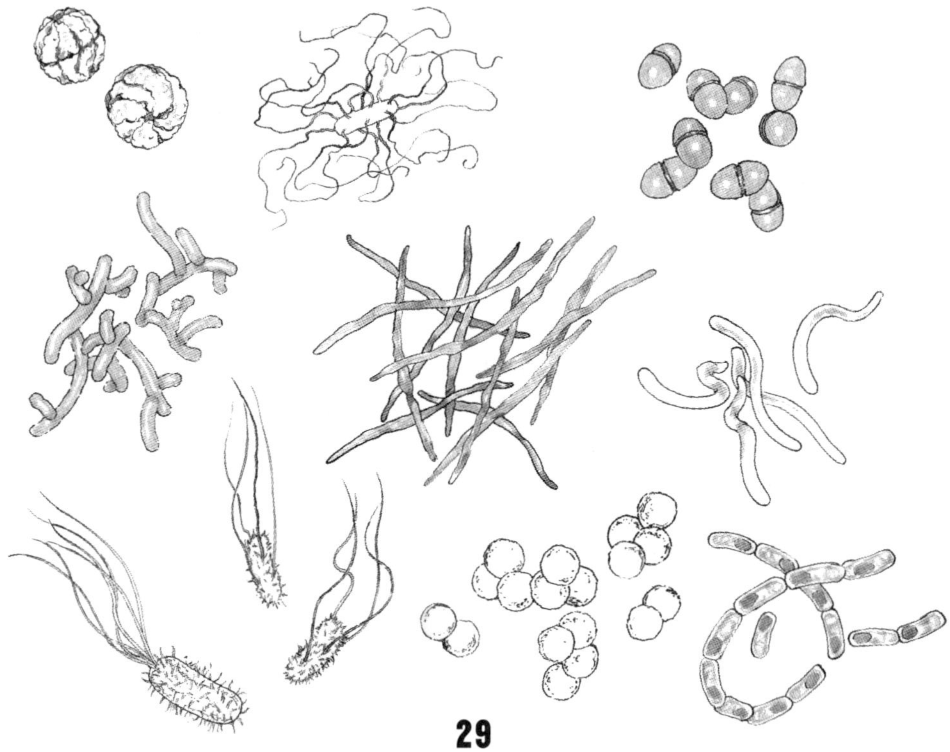

others look like rods. Some are capable of motion; others are not. Some thrive in cooler temperatures and others require heat. The specific aerobic bacteria that will be present in your pile will change continually, based on factors like temperature, moisture, and what materials you've added.

### Actinobacteria

Actinobacteria are fungi-like bacteria that are also essential in the breakdown of organic matter. They are more complex than single-cell aerobic bacteria and function similarly to fungi in some cases, but they form their own distinct group. They are largely responsible for the "earthy" smell we associate with good soil. In a mature pile that has been sitting for a long period of time, actinobacteria can become visible to the eye as white-gray patches of powder. They are excellent for breaking down hardier substances, like lignin found in wood and cellulose found in grasses and other plants.

### Fungi

Fungi are a huge part of the decomposition process. They show up a little slower than bacteria but are more skilled at breaking down the recalcitrant materials left behind by feasting bacteria. The majority of the fungi in your pile will be saprophytes, which specialize in breaking down the complex carbon-based molecules in dead or dying plants and animals. Fungi will slowly form visible networks of mycelium, and you might even see actual mushrooms emerging from your compost.

### Nematodes

Nematodes are very small roundworms. Some eat bacteria, some consume fungi, and others will eat each other. In doing so, they continually free nutrients that have been locked away in the bodies of other microbes, excreting them back into the pile where they are available for continued foraging and plant roots. Some of these guys also break down organic matter in the same way as worms, but at a much smaller scale.

### Protozoa

Protozoa are one-celled organisms that include flagellates, amoebae, and ciliates. Their primary role in the compost pile is to eat everything else. They consume organic matter directly, like bacteria and fungi do, but they also eat bacteria and fungi. By doing so, they play a key role in regulating your pile's population of aerobic bacteria and ensure the continued cycling of nutrients. Eat and be eaten, though. Protozoa are a key food source for worms.

### Other small creatures

As we move into the realm of the visible, decomposers take on more familiar forms. Here we have worms, beetles, millipedes, sow bugs, spiders, and grubs. These guys feed on decaying matter, microbes, and each other. They are also the "shredders and movers" of organic matter, transforming bulkier materials into smaller pieces and distributing them throughout the pile.

BLACK BEETLE

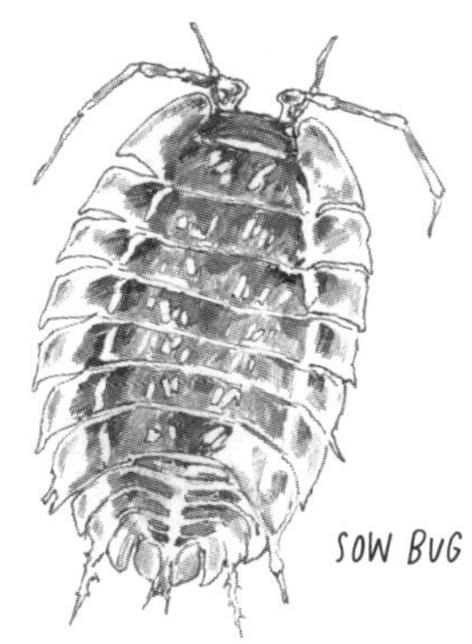
SOW BUG

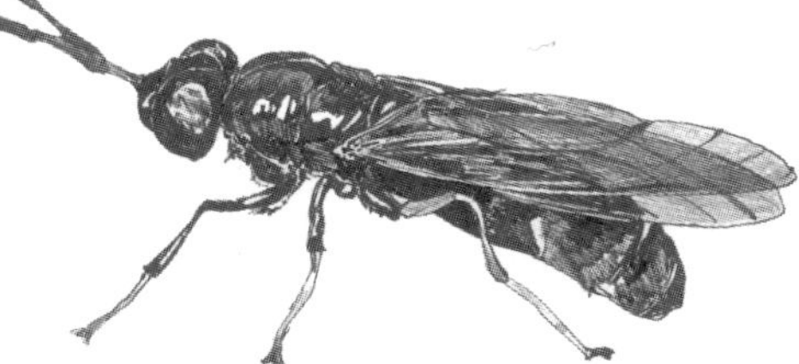
BLACK SOLDIER FLY

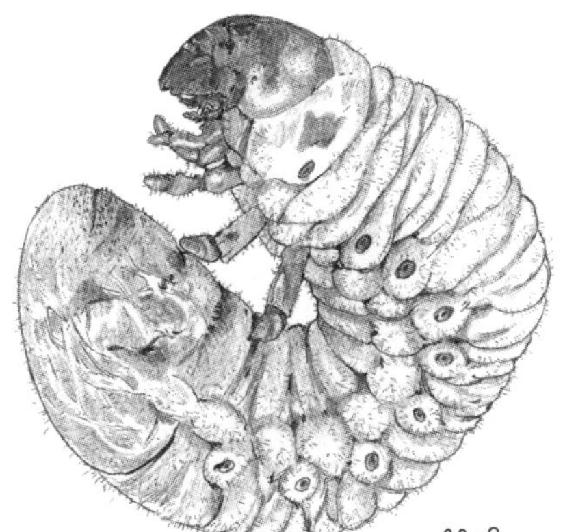
GRUB

PILL BUG

MILLIPEDE

You may also find snakes, lizards, frogs, and other small creatures seeking refuge in the warmth and security of your pile. The presence of these creatures is good. They mean your pile is a home, able to provide the abundance that other things require to live, which is a sign you're doing things right.

# THE MYTHOLOGY OF "AWAY"

A friend of mine grew up on an island in the Pacific Ocean off the coast of Canada. It's a small island with several farms, dozens of chickens, and some goats. There is a post office that is served each Thursday by boat. Not much else. In fact, trying to describe the island feels more like a list of what they *don't have* versus what they do. There are no stores and no restaurants. No hotels or pharmacies.

Things that go there tend to stay there.

It follows that there is no trash service, then, and absolutely no dump. In fact, there is really nowhere for anything to go that isn't right in your face. Trees are felled and chopped, and any pieces not used for kindling are piled up and left to decompose. Weeds pulled from crops are heaped in place to be utilized as mulch. Goats that died were piled into a pit layered with biochar and covered with earth. Many of the homes are even equipped with composting toilets.

Every single person on the island, consciously or not, composts almost continually. In fact, the entire life of the island flows along the contours of its relationship to waste. People opt for materials that biodegrade and adopt habits that do not contribute to material accumulation. As a result, they live within the bounds of their own capacity for disposal, accruing and declining in equal measure, never more.

It might be difficult for some of us to imagine that way of life. Right now, so much of the way we live is propped up on the great lie that is "garbage." We purchase freely and throw away with abandon. Every week, we load up our trash cans and cart them to the curb, where they vanish into the maw of a truck and are whisked away forever.

There is no such place as "away," though.

ISLAND COMPOST PILES

THE AUTHOR TENDS HER COMPOST.

The garbage always goes somewhere. It's stacked in landfills, shipped to other countries, and dumped into the wild. The Great Pacific Garbage Patch, a mess of plastic waste that has accumulated in the Pacific Ocean somewhere between California and Hawaii, is estimated to be twice the size of Texas. In Accra, Ghana, tens of millions of clothing pieces are imported each week—most from fast-fashion companies who advertise this service as a "charitable donation." Nearly half of that clothing ends up as litter. It rots in towering heaps around the city and along the shoreline, where it is grazed heavily by animals or dug into informal trenches, polluting the surrounding soil and waterways.

Formal landfills are not much better. Even when preventative measures are taken, they leak contaminants and emit voluminous amounts of noxious gas. This makes them an enormous contributor to global pollution. According to a 2023 EPA report on food waste, landfills in the United States release hundreds of millions of tons of methane each year, with 58 percent of those emissions coming just from rotting food waste.

There is some irony in the fact that people find compost "gross." They worry it will smell bad or attract animals. It seems time-consuming and inconvenient. Time and again, I hear from people that it's cleaner to just throw things in the trash. It's not. The difference between the "grossness" of compost and that of your garbage can is only a matter of deniability. Your garbage gets carted "away," allowing you to ignore the consequences of your consumption. Compost does not permit this delusion. Instead, it invites you to be accountable.

Learning to compost radically shifted my relationship to both garbage and my own consumer habits. These days, almost nothing in my house goes in the trash. And what I can't compost, I avoid buying. I compost my hair, nails, blood, dryer lint, dog's fur, and leftover tater tots. I compost my old cotton clothing when it's too ripped up to wear anymore. I compost my dead plants, my cardboard boxes, and sometimes my junk mail. Once, I even composted an old bible I found. All of it goes into the pile and disappears. I find this amazing and also quite fun, like a never-ending science problem that I can playfully solve with my own two hands and some tinkering.

I've witnessed similar transformations in others. Like a modern-day alchemist, the individual who learns to compost comes to delight in their near-magical ability to transform worthless matter into something entirely new and vital. Their perspective adjusts to suit the procedure. Things made from plastic suddenly seem so strange. Buying less overall becomes a given. Once the compost pile has provided a person with both the ability and curiosity to manage their own waste, they suddenly become inspired to live within the bounds of their own capacity, just like my friend on the island.

A SIMPLE GARDEN COMPOST PILE DESIGNED FOR MY FRIEND DAVID

# THE BEGINNER'S MINDSET

The most helpful thing I did when first learning to compost was to embrace the fact that I had no idea what I was doing. This is the basic premise of what is called "a beginner's mindset." Not feeling like I needed to know everything freed me from the pressure of perfection, and it allowed me to give myself permission to make—and learn from—mistakes. It's been years now, but no matter how much I learn about compost, I still enjoy thinking of myself as a beginner.

Being a beginner is just more fun. It's casual and allows for experimentation and creative outbursts. Some composters I know have been running their piles for decades and still comfortably assert that they don't really know anything about compost. I once met a quite well-known and respected composter at an event in Ohio. "The more I learn about compost, the more I realize I don't know," she told me. Exactly! This kind of attitude allows you to remain permanently open to learning. I love that.

It also works particularly well with composting because, to a certain degree, compost is a fundamentally unknowable thing. Its processes, as varied and overlapping as they are, remain at least a little mysterious to even the most accredited expert.

As beginners, we embrace this. We can know what is happening in our compost intuitively and even experience it physically through what we can touch and the things that we smell, but there are limits to what we will ever quantify scientifically.

The other features of a beginner's mindset that will serve you well are trust and humility. Trust that you will learn as you go, that your observations will be sound and your intuition accurate. As you practice, you will gain confidence. You will get things right over time, and you will learn from anything you do wrong. Trust yourself and trust in the resilience of your pile. Humility helps you welcome the idea that, even when you think you've got it down to a "science," the mysteries of compost will continue to confound, propel, and excite.

At the end of the day, the idea of "composting" includes an enormous number of possible needs and outcomes. Every pile has a remarkable ability to absorb errors and to express its own preferences to the observant and thoughtful caretaker. You can and should build your relationship with your heap in order to develop your own passionate and illuminating commitment.

Here are some basic principles that can guide you.

## Don't hide your compost.

In order to learn, you need to observe and interact. What do you smell? How do things feel? Is your pile wet to the touch, or is it too dry? Are things breaking down or lingering? These simple and observable characteristics will help you make small adjustments to what you add to your pile and how often, in addition to other forms of maintenance. In return, you'll end up with healthy and efficient compost that looks good and is virtually odorless.

- **DECOMPOSITION IS DESTINY.** Every biological material on Earth will eventually decompose. No matter what biological materials you add, what order you add them in, how often you turn things or don't, everything will decompose. There is absolutely nothing you can do to make that *not* happen. Let this embolden you to begin.

- **KNOW THY PILE, KNOW THYSELF.** Your work is to develop a relationship to your unique pile, so that it meets your specific needs. Do you need your compost for a garden? Then you may want to make sure things decompose on a timeline that suits your planting schedule. Are you worried about smells because you have neighbors nearby? Then you'll want to be fastidious about keeping your compost maintained—and potentially covered—in order to minimize odors. Are bears a problem in your area? Then make sure to keep your compost far from the house. Do you just want to get rid of your food scraps in an environmentally friendly way and don't need the end product for anything specific? Then you can do pretty much whatever you want.
- **RELAX.** Spreadsheets, precise ratios, and measurements are all exceptional tools for the composter, but they are not necessary to compost well. You are allowed to compost however you see fit and in whatever way you are capable of doing so. If somebody tells you a "rule" about composting that contradicts your observed experience of your own pile, feel free to disregard it.

# WHERE TO PUT YOUR COMPOST

Questions of location are important to consider. Should your compost pile be put in the shade? What about the sun? Does it need a lot of space? What about animals? These are all quite common questions, but they neglect to address the single most important factor regarding the success of any given compost heap: *you*. As the composter, your own physical abilities and daily routines should be the first point of consideration when determining the ideal location for your pile.

I can't tell you the number of composts I've "fixed" through simple relocation, making the heap more accessible to the daily routine of the composter. One woman I received a phone call from had her heap boxed in with wooden fencing on the far side of her rather generously sized property. "I know I love to compost, but I just don't ever seem to add my food scraps to the pile," she lamented, as we made the five-minute walk from her house to where her heap resided. To me, the issue was immediately clear. If she wanted to develop a relationship with her compost, her pile needed to be closer to the place where she spent her time cooking, so it would be much easier for her to add her food scraps on a regular basis. I gathered her compost and moved it into a simple stone enclosure just outside her back door. Her compost practice flourished soon after.

Another friend once told me his compost was "broken" and asked if I could check it out. When I arrived on his property, I found that he'd piled his heap into a very tall and narrow chicken-wire enclosure, making turning a clumsy and onerous procedure. As a result, he had begun to neglect the pile, leaving it dried out and inactive; decomposition had stopped almost completely. I cut open the chicken wire and rerouted it, adding about 3 feet in diameter to the heap. A little more space made turning the pile much easier, which invited him to interact with it more often. As a result, he turned more frequently and kept

things more consistently moist. Soon, his compost was humming along at a neat and active 110°F.

I often encounter this problem of "mysterious" neglect when people build their pile in out-of-the-way places or with cumbersome setups that make basic interaction a challenge. Simply put, your pile will thrive if you build it in a place where you can run into it on a daily basis and in a way that aligns its maintenance with your abilities. That's why I joke with people that making your first compost is a little like going to therapy. Making it work requires you to look deep within yourself. You need to understand your own habits, proclivities, and limitations.

In general, having your compost pile located in a place that you regularly see acts as a reminder to *care* about it. It also helps you learn. Your day-to-day observations will teach you all kinds of things about your heap, like when things are looking dry and need moisture, or when you may need to adjust the balance of your inputs because of odors. You'll be amazed at how quickly you pick up on the "language" of your pile just by looking at it every day.

Of course, not everyone has access to so much space that their own preferences and habits get to determine the location of their compost. Living in cities or with nearby neighbors means taking other people's needs into account too. For example, if your ideal compost location ends up being directly beneath the bedroom window of a neighbor, you might want to rethink your approach. (Or you could brave a conversation and ask for permission. You never know, right? You might make a new compost pal.) No matter what space you have available, though, just do your best. Wherever your compost ends up, you will be able to make it work.

# HOW TO PICK YOUR COMPOST SETUP

There are so many ways to build a compost pile. Each pile will be shaped by the composter's specific circumstances, aesthetic preferences, physical ability, and degree of interest. Whether you have land or not is a big factor. Whether or not you're an artist or a parent—or both—or have pets, or cold winters, or bears, or eight full hours a day of unrelenting and precious sunshine can all be important considerations. Some people don't really care about their pile and can't believe when their compost works anyway, and some people are meticulous with measurements and take temperatures daily. The point is, your compost setup is up to you to decide, and it doesn't need to be based on anything more than who you are and what you need. Here are some questions you could ask yourself to figure out how you want to approach things: How much space do I have? How much time do I have? Do I care what it looks like?

Your answers to some questions will be determined by need, while others will be guided by preference. Some will overlap and interact with each other. You can consider them separately, but you will likely need to also consider them in relation to one another. Being aware of them will help you arrive at a compost setup that works for you, reflects your needs, and will be conducive to building a strong and rewarding composting habit.

### How much space do you have?

This will likely be the strongest determining factor in choosing your setup. All other factors can be flexible, but space rarely is. A small apartment, a house with a tiny yard, and a ranch house with full-blown acreage will each accommodate vastly different approaches to composting. The apartment is the most constricting, and you'll probably want to explore a one-bag, under-the-sink setup or an alternative like bokashi. (I'll discuss these options in more detail later.) If you have a yard, you have more options and you can defer to your preferred time commitment and level of effort as determining factors. A small, cemented back patio is another common area that people have and can accommodate a wide range of compost approaches, from one-box composting to prepurchased bins.

### How much time do you have?

Many home composters are busy people. If you have limited time or are uncertain about how much active maintenance you desire for your pile, your best bet is a single pile structure with a lid. The single pile system accommodates an ebb and flow of attention without major issue, easily able to withstand periods of neglect. If you have lots of time to devote to your pile or want to work up a sweat, you can build out a more complex structure that takes some regular maintenance. I would recommend a traditional three-bin system. (Again, more on all this later.)

### Do you care what it looks like?

If you're sensitive to how your compost will look in your yard, there are a variety of nice-looking bins that you can purchase at hardware and garden stores. Look for a bin with a lid and a "door" at the base that you can open to retrieve the finished compost. If you don't particularly care what your compost looks like, or you find the idea of a compost pile to be generally aligned with your existing aesthetic preferences, you get to have a little more fun. You can have an open pile or DIY a structure from found materials, like cinder blocks and chicken wire. The only limit, really, is your imagination. In general, I don't recommend any compost setup that is too expensive or promises an overly technological solution. These products, in my experience, never improve enough on the efficiency of a regular pile to merit their price tags.

## Why I don't recommend compost tumblers

I don't know many people who have good luck with compost tumblers. They're great in theory: a tidy, aesthetically pleasing solution that promises to hide your compost from view while protecting you from odors and animals. However, all the things that make tumblers so appealing to the amateur composter are all reasons—in my opinion—why they fail. When you never see or interact with your compost, you are less likely to develop a fluency with the way it decomposes, which means you are more likely to end up with all the problems you tried to avoid by purchasing the tumbler in the first place. Tumblers tend to get overstuffed, leading to compaction, and they generally don't facilitate good airflow, which can cause problems for the breakdown of matter. They're more likely to develop odors, because people can't easily eyeball the total ratio of their materials. Worst of all, they tend to get forgotten about altogether. Most people I know struggle to turn a tumbler purchase into a full-blown composting routine.

# WHERE TO SOURCE YOUR COMPOST MATERIALS

Before building your first compost pile, it can be good to assess what types of material you have around. What does your household waste look like? Think about your weekly habits. What are you throwing away and how often? Is the majority of your organic waste composed of materials that are high in

nitrogen, like food scraps or coffee grounds, or do you mostly deal with high-carbon stuff, perhaps landscaping castoffs or a lot of cardboard boxes from deliveries? Crucially, do you generate waste that offers a good mix of both types of material?

Not everybody does. In fact, it's quite common that people have an excess of one and not much of the other. In those cases, you can get scrappy and source materials that will help you maintain a good balance. If you need more nitrogen, ask your local coffee shop for a bag of their used coffee grounds. If what you lack is carbon, make friends with the tree trimmer on your block and ask if you can bag up some of their wood chips

C C C C C C C C C C C C C C C N

EGGPLANTS & TOMATOES, 15:1 CARBON TO NITROGEN

DRIED OAK LEAVES, 60:1 CARBON TO NITROGEN

to bring home. Or the next time you're at a friend's house and they're about to toss some cardboard into the recycling bin, intercept it and take it home. You can also purchase materials, if you're in a pinch. A box of alfalfa is a great source of nitrogen, and it doesn't cost much.

Over time, you might come to find that curiosity about the material world fills you with delight. Objects in your routine are no longer disconnected entities, available to you just for aesthetic pleasure or brief utility. Instead, they practically vibrate with potential life. Garbage becomes especially fascinating. Another person might throw out that load of wet cardboard, but you can do something tremendous by transforming it into fertile earth.

Nurturing your curiosity about the world of materials around you in this way will have a tremendous impact on your composting practice. I encourage spontaneous and continuous investigation. Start researching the things in your house. Think about what books are made of, and also your dryer lint. With practice, you'll develop some fluency around what different things are made from. In fact, you may even find yourself subconsciously dissecting the components of every item you encounter.

These days, I get a lot of very funny phone calls from people who are looking for better things to do with their garbage than sending it to a landfill. I once received a text from the headquarters of a clothing brand that had left a shipment of empty cardboard boxes in the rain; would I be interested? I nearly jumped out of my seat. That cardboard, a pre-moistened source of microbe-nutritious carbon, soon ended up in compost piles at a succession of community gardens around Los

Angeles. Another time, a friend messaged me about an artist's exhibition he had just attended. "There is a huge sculpture made entirely out of manure," he wrote. "Would you be interested?" And of course, I was. There is no need to look far for things to add to your compost pile; many materials are great for composting and can easily be found almost anywhere.

### Food scraps

Food scraps are the most regular source of nitrogen for most people's backyard compost piles. This includes bread, citrus peels, eggshells, and even cooked leftovers. However, you should be aware that cooked food tends to be aromatic enough to attract animals. Some people opt out of adding it to their piles for that reason.

### Cardboard boxes

The all-natural fibers of a regular cardboard box make for a great high-carbon addition to your compost pile. Make sure to peel or cut away any areas that have adhesive on them, then shred the box before adding it to your pile. Cardboard is sturdy enough that it can help sustain airflow throughout your pile. It can also help suppress odors if you layer it on top. However, it is important to note that not *all* cardboard boxes are compostable. I would avoid composting any that are glossy or feel coated. These coatings are often made with synthetic materials and can be printed over with synthetic inks. Porous, uncoated cardboard is generally safe to compost and tends to be printed with bio-based inks. Here's a good rule of thumb: if a cardboard box seems too brightly colored or too shiny to possibly be "natural," you should probably trust your instincts.

### Paper towels

Paper towels are a great high-carbon addition to your pile. You can compost white paper towels, brown paper towels, paper towels made from bamboo—you name it. You can even compost them if they're coated in food grease or other residues. However, avoid composting them if you've used them with household cleaners or any other chemicals of that nature.

### Incense ash

Incense ash is fine to compost, particularly since it would tend to be available in such small quantities. It will add a nice dose of micronutrients to your pile, particularly potassium.

### Flowers

Not only can you compost flowers, you *should*. They make your pile look quite pretty, like an art sculpture. I work with my local farmers market to compost any organic leftovers, which includes whatever flowers don't sell that day. I always add the flower remains last, "closing" the pile by laying their fragrant blossoms over the top. Composting your cut flowers feels like it should be a requirement to honor their natural life cycle. It may be worth noting that some floral arrangements utilized for events or decoration are treated with a spray or sealant to improve their longevity. You should avoid composting blooms of this variety. Remember to ask if you're unsure!

### Cardboard tubes from toilet paper

Both toilet paper and their tubes are compostable. Save them, shred them, and add them to your pile. Once you do, I assure you the act of "recycling" them in the bin will feel practically boring.

### Tampons

Yes, used tampons can go into your compost pile. The tissue from your body is natural. Your blood is natural. The all-cotton fabric of the tampon is natural. No reason to throw a tampon in the trash. Return it to the earth.

### Printer paper

As long as your printer paper is not glossy, you are good to add it to your compost. I printed out drafts of this book on recycled printer paper, edited by hand, and then shredded and composted the remains once I had codified all the changes on my computer.

### Mail

I once composted my tax documentation instead of shredding it and throwing it in the recycling. For some reason, it was incredibly satisfying. You don't want to compost junk mail that arrives on glossy paper or stuff that's coated with unknowable, probably plastic substances. My general rule is: Does it feel smooth? Throw it in the trash. Does it feel rough and porous under the fingers? Probably okay to compost.

### Newspapers

Newspapers are compostable and a great source of carbon for your pile. Shred and wet them before you add them in order to hasten their decomposition.

### Grass clippings

Green grass is a great source of nitrogen. Add it in thin layers throughout your heap, making sure to mix it with bulky and high-carbon materials to prevent compaction. There are mixed ideas about what to do with your grass if you treat it with chemicals. Some say don't compost in those cases, while others think that a consistently hot pile will prevent any chemical contamination. I'm of the mind that you should stop chemically treating your grass altogether, rather than worrying about whether those chemicals will harm your compost.

### Leaves

Leaf litter is the planet's natural compost, carpeting forests and hillsides, decomposing in place and providing the soil with an abundance of rich nutrients. They are also a great high-carbon addition to your compost pile, making them a staple ingredient for the backyard composter. Leaves that have fallen in your yard are excellent to collect and add to your compost. They will bring with them their own population of good little decomposer microbes, which should enhance the overall pile. They may take a long time to break down if you don't shred them before adding, however. You can use your hands to rip them up, run them over with your lawn mower, or invest in a commercial shredder—all are good options.

### Straw

Straw works best in a compost pile if you add it after it has been weathered, which means left out in the elements until it starts to break down on its own. This is because it is such a high-carbon, sturdily constructed material; it can require a lot of nitrogen to break down properly in a pile, so allowing time for decomposition to kickstart before adding it in is helpful. I've had great success with straw that has been used as bedding in chicken coops. The animals' droppings are very high in nitrogen and balance the carbon in the straw well, making for very prodigious decay.

### Tea

Loose-leaf teas are a great addition to your heap. Proceed with caution when composting tea bags, however, as sometimes they contain plastics. I always like to check the box and see if the material the tea bag is made of is listed.

### Pine needles

Pine needles are compostable, but they break down very slowly. As a result, they are best suited for adding in small quantities or when broken down into smaller pieces first. They are quite acidic when freshly fallen, so if you are composting them in great quantities, you may want to balance their addition deliberately with materials that are alkaline, like eggshells.

### Coffee grounds

Coffee grounds are a high-nitrogen staple for any compost pile. My home compost sometimes feels like it is made entirely of coffee grounds. As I mentioned before, going to your local coffee place and asking for their old grounds is a great option to consider.

### Oyster shells

Mollusk shells are very, very difficult to break down. I have a "lucky oyster shell" that I leave in the compost at the garden where I work, just to see how long it takes for any visible sign of decomposition to emerge. It's been there for four years so far. Nothing about it has changed. The game we play is that whoever finds it on any given day gets good luck for the next twenty-four hours. If you want to add shells to your pile, grind them to a powder beforehand. They are a great source of many minerals and can help balance pH; they just require a bit more labor.

## Things you may have been told you can't compost (that you absolutely can)

Everyone has their own rules about what can be composted and what you should never put in your pile. However, I find these rules are rarely applicable across the board. There may be good reasons why your neighbor won't compost eggshells or bread, but that doesn't mean *you* shouldn't feel free to.

### CITRUS PEELS

Citrus peels are a great source of nitrogen, which every heap needs. They also offer micronutrients like phosphorus and potassium, which your future plants—if you use your compost to garden—will love. Whether or not you can compost citrus peels is really a matter of proportion and personal preference, though. For example, adding loads of lemon peels at once may tilt your compost more toward acidity or attract fruit flies. You can avoid these potential consequences by adding peels in balanced proportions with other types of food scraps, covering them with a bulky carbon material like dead leaves, and chopping them into small bits prior to adding.

### BREAD

Bread, being composed entirely of organic materials, is eminently compostable. Some gardeners suggest that you shouldn't include it because it has a higher risk of attracting animals than other types of food scraps. The risk is yours to determine, but you should know that it is absolutely fine to compost bread—even if it's moldy.

### EGGSHELLS

Eggshells are a great source of calcium and can help adjust overall pH if your compost is on the acidic side. They do take a while to break down if you add them without crushing or grinding them first, though. My advice is to crunch them up in your palm before adding, or they may last forever in your pile.

### MOLDY BITS

Mold is a natural part of the decomposition process, like the other microbes that are already present in your pile. If you have moldy food—whether that's fruits and vegetables or a hunk of bread—you can feel completely confident adding it to your pile. It will be right at home there.

### URINE

Urine is fine for your compost and, being high in nitrogen, can even act as a handy activator for decomposition. Urine can be good for your pile if the pile is heavy in high-carbon materials, like it may be after you add an abundance of leaves after raking the yard. It can also be good if you're just starting your pile and trying to get things going. I have more than once advised friends to pee on their piles for this purpose.

### HAIR

You can compost hair, although it may take a while to break down if you add it in large amounts. I compost my hair after haircuts and my dog's fur after a good brushing. Both are fine. I read once about a man in Texas who devised a compost recipe made almost entirely of human hair and intended for a special breed of rose. This implies that not only is it okay to compost hair, you could theoretically build an entire pile out of it. Ambitious.

### NAILS

Nails are compostable. In fact, the next time you need to clip your nails, try doing it directly over your pile. Might as well not waste the nutrients, and it'll give you a chance to check in with your pile at the same time.

### WEEDS

You can absolutely compost weeds. Weeds are just plants, like any other. However, if you don't want to have to pluck sprouts from your finished compost, it'd be best to only compost them if you have a consistently hot pile. Personally, I don't mind the additional maintenance that may be required if I am gardening with compost that has some weed seeds in it. All kinds of plants are always popping up in my beds for all kinds of reasons anyway. Some I have to get rid of, and others get to stay. It's a part of gardening, and a few seeds in my compost aren't going to make a meaningful difference. You are best off not composting diseased plants, though. Diseases, unlike weed seeds, cannot always be easily managed later if they survive the composting process.

### ONIONS

It's been suggested that you shouldn't add onions to a compost pile because they smell particularly bad, and it's said that their acidity can have a detrimental impact on the microbes in your pile. My feeling is that your microbes are quite resilient and will not be enduringly harmed by having to digest a couple onions. If their numbers are briefly suppressed, they will recover as you continue to add more and different types of material to the pile. Perhaps don't make a compost entirely of onions, though.

### MEAT

You can compost meat. That doesn't mean you should, of course. Adding meat to your compost bin will significantly increase the likelihood that it will attract animals. This can be a big problem if you live in an area with bears. It can also be problematic if you have neighbors and they don't particularly like rats. If you do opt to compost meat, you may want to bury it in a (deep) hole and then cover it.

### OILS

You can compost leftover food, even if it has been cooked in oil. I have heard from people who think they shouldn't compost oils because they can slow down the overall pace of decomposition in the pile. While this might be true if you compost a 5-gallon bucket of olive oil, I don't think the dregs of your pasta dish will meaningfully impact your pile's rate of breakdown.

# HOW TO BUILD A PILE

No matter what type of compost pile you aim to build, where you plan to put it, what kind of place you live in, or what type of person you are—the basic composition and needs of your compost pile are going to be the same.

Your compost will need a balance of materials high in carbon and materials high in nitrogen, at roughly a 2-to-1 ratio. It will also need consistent airflow and moisture. It is your job, as the composter, to construct your pile from these pieces and then work to maintain them over time. You don't need to be punishingly exact. Instead, just keep these general principles in mind and let them guide you as you build and nourish your first pile:

Balance. Moisture. Airflow.

You can repeat it to yourself like a little compost mantra.

- **BALANCE.** Your balance will come from maintaining a roughly 2-to-1 ratio of high-carbon to high-nitrogen materials. You can build your first pile using this ratio and then add new materials in roughly the same proportions.
- **MOISTURE.** Your moisture content will come from some of your materials themselves, as they break down. You will also need to add water to your pile with a hose from time to time. Rainfall and snow can also add moisture to a pile.
- **AIRFLOW.** Turning your compost will create airflow, as will the incorporation of bulkier and sturdier high-carbon materials, like thick wood chips and twigs, throughout the pile. The type of compost enclosure you choose should accommodate at least some aeration.

Now, let's build your first compost pile. Don't worry about measuring anything out in highly precise quantities. Just approximate your portions and be ready to observe what happens and anticipate future adjustments.

Here's a simple recipe you can follow:

1. **START WITH A LAYER OF HIGH-CARBON MATERIALS,** like broken twigs, shredded cardboard, or dry leaves. These will be the base of your pile. They will help oxygen flow from the bottom, where things are more likely to get compacted as your pile grows.

2. **NEXT, ADD A LAYER OF HIGH-NITROGEN MATERIALS.** When you're done, add more high-carbon materials, ensuring that they fully cover the previous layer. Add enough water to dampen things.

3. **STEP BACK AND TAKE STOCK.** Are your high-nitrogen materials fully covered? Are your high-carbon materials fully moistened? Run your hand over the top layer and see if the water you added has penetrated beyond the surface. If it hasn't, give things another shot of water. Things have begun.

4. **KEEP LAYERING YOUR PILE** using that same rhythm until you've added all the materials you have on hand: more high-nitrogen material, then more high-carbon material. You may want to add a handful of twigs or thicker branches to the middle portion of the pile to ensure proper aeration. Continue adding water intermittently. You want the pile to be moist throughout but not sopping wet.

5. **FINISH THE PILE WITH A THICK LAYER OF HIGH-CARBON MATERIAL ON TOP.** This could be a bunch of leaves or sawdust, or even a layer of flattened cardboard. This will help

suppress any odors, in addition to deterring animals and insect infestations. It also, frankly, looks a little bit nicer than exposed food scraps.

**6. THEN, JUST LEAVE THINGS ALONE.** The final ingredient of any good compost pile is, simply put, a little bit of patience.

# DEMYSTIFYING TEMPERATURE

Compost can get hot, but it is a common misconception that this heat is a result of exposure to the sun or some other external factor. In fact, the heat is generated from activity within the pile itself. As microbes break down organic matter, they turn it into fuel for their own cellular processes, which creates heat energy. In a decently sized pile, that heat builds up. You'll notice that your pile becomes hot to the touch, and you might even see steam rising off it.

In composting, there are two main temperature ranges to consider: mesophilic and thermophilic. A mesophilic pile is between 50°F and 113°F. This temperature range accommodates the greatest diversity of microbial life, with aerobic bacteria, actinobacteria, and fungi all multiplying rapidly. Many pathogens are also able to thrive, although they compete with other microbes for energy sources and don't always survive as a result. A thermophilic pile is between 113°F and 158°F. At these temperatures, aerobic thermophilic bacteria flourish. Pathogens die off, but the overall microbial life is also reduced in diversity. Thermophilic decomposition is quick, and when a pile is in this range, things break down rapidly. That's one of the reasons many gardeners aspire to a thermophilic pile. It makes for efficient and rapid decay, while minimizing the risk of contamination.

Compost piles will only stay in the thermophilic range for a limited amount of time. Once it isn't possible to break the material down any further, thermophilic microbes will begin to die off, the heat energy they created will ebb, and mesophilic microbes will regain their dominance as the pile cools. The dropping temperature is a good signal that the compost in your pile is nearing completion.

Piles can get hotter than 158°F, of course. While it can be tempting to see how hot you can get things, "hot" is not necessarily good. High temperatures have less and less microbial diversity, and less diversity equates to a less nutritious final compost. You're also at greater risk of scarier things, like combustion within the pile.

Most backyard composts will fluctuate between the mesophilic and thermophilic ranges as new material is added, the pile is turned, and its size grows and shrinks. This is normal and to be expected.

If you want your compost to be hotter and can't figure out why things aren't warming up, you may want to double-check the key features that contribute to heat potential: the overall size of your pile and the types of materials you are adding. Common compost wisdom dictates that a pile needs to be around 3 cubic feet to successfully retain enough of its own heat to steam up. If you add too many materials that are high in carbon without balancing them with enough nitrogen, you will also struggle to get a hot pile. Your microbes just won't be able to break stuff down fast enough to multiply at the rates required to achieve high temperatures. Try shredding things down into smaller pieces, adding more nitrogen if you need it, and heaping your pile up to at least 3 feet high. Give it a week, then see if it's hot to the touch.

One of my favorite letters I've ever gotten was from an enthusiastic and kind composter located in northern Florida. He wanted to ask me about ideal airflow because he was certain imperfect aeration was the reason his own pile wasn't reaching high temperatures. He had rigged his bin up with several tubes that would pump air throughout the pile at varying rates, and he felt he had tried everything but it still wasn't working. After thinking about it for a minute, I wrote back and asked what was *in* his pile. It turned out the answer was mostly oak leaves. Bingo. The problem was not how much airflow the pile was receiving. It was simply that oak leaves are a very sturdy, high-carbon material that is difficult to break down. I told him to shred the leaves, balance them with some high-nitrogen inputs, and see where he landed.

Even if you do get your pile up to temperature, though, it will always benefit from a period of rest once the bulk of its materials have fully decomposed. This is a maturation period that allows the microbial community to rebuild and diversify, which enriches the final product. Just remember that even a resting pile will still need at least a little care—specifically, you want to keep things moist. Microbes need that water to sustain and flourish.

### In defense of cool compost

Although compost *can* get hot, not all composts *have* to be hot. Personally, I prefer composting at cooler temperatures. The pace of decomposition is slower, but a cooler pile supports a greater diversity of lifeforms, from aerobic bacteria and protozoa to worms, grubs, and even the occasional frog.

A cooler temperature is best achieved with a single pile where you add things erratically and turn infrequently. Coincidentally, I've found this to be the type of compost that also works for the broadest range of people. It requires you to operate on no particular schedule and to do as little work as possible. It also doesn't require much space.

My home only has a tiny back deck. It's not big enough to host a compost bin, but I make do with a wedge of space between the deck and my kitchen window. The pile itself is built from food scraps, paper towel bits, dog fur, nail clippings, and whatever else is biodegradable and within reach. It never has a reliable shape or temperature. It grows and shrinks and grows again. It stays mostly cool and works very slowly but, over the years, I've found it to be both alive and generous, full of earthly surprises. For example, if I add anything that used to be food, it eventually erupts with renewed and vigorous life. Zucchini seedlings grow from the desiccated bodies of old zucchini, carrot tops become leafy with new growth, and withered radishes begin to sprout thick and twisted roots. Critters also flourish. A population of black beetles ebbs and flows, and earthworms are always present. One time, after heavy rains, I found a California slender salamander tucked tenderly beneath some partially decomposed leaves, as slight and shy as his name would indicate. He stayed completely still as I gazed at him, only betraying his life with a set of slow and gentle blinks.

These types of experiences would simply not be possible if my compost pile was consistently hot and, to be candid, managed more efficiently. For compost, as with other things, efficiency can be valuable, but it can also be a form of constraint. It does not allow for the unexpected occurrence. My cool compost may be haphazard, but it offers me the delight that only a surprise can engender.

My pile has also been a great and steady teacher. It offers patience in the face of urgency, an awareness of the time it takes for things to change. Sometimes, when the problems of the world feel gigantic and I feel very small, it is nice to have a practice that codifies these principles. When no big gestures can be made, what's left? The smallest. The unintentional home for a salamander. The vitality in which other things can grow.

# LEARN BY DOING

With the right amount of attention, your pile will teach you what it needs to work. It will teach you better than most books can, even this one. And it will teach you through what you get wrong, in addition to what you do right. In fact, I would venture to say that your practice will be stronger—and your knowledge deeper—if you give yourself permission to risk doing things wrong, trusting yourself to learn from your mistakes. Compost is nothing if not resilient, and it will sort itself out if you add too many food scraps at once, or don't touch it for six months, or turn it on Sunday instead of Wednesday.

The best possible thing that you can do is get started. Give yourself something physical to interact with, observe, and learn from. This will sharpen your expertise over time, but it will also make your expertise specific. It's not just "compost" you are learning how to make. It's *your* compost, and your compost will always be at least a little unique to you and your circumstances and needs.

It will take some time for your pile to build up the microbial population that makes for active decomposition, which means you may not notice things happening right away. That's perfectly normal. Just keep adding new materials to the pile as they accumulate, using those additions as an opportunity to observe, touch, and smell. Casually take notice of things like how broken down stuff is or isn't yet. Assess the overall size of the pile. Check if there are any visible creatures beginning to appear, like grubs or beetles. Check if things are staying moist throughout the pile. Add water if you notice it's dry.

Every time you add new material, you'll work to maintain your pile's overall balance, moisture, and airflow. Every time you add food scraps, make sure to add some high-carbon material as well. For example, if you're adding a big bowl of food scraps from last night's dinner, throw in a few handfuls of dead leaves at the same time. If you're low on high-carbon materials, you can dig a small hole in the top of your pile for your addition and then cover it up again, keeping any food scraps sufficiently covered.

After a few weeks, you're likely to notice that things are starting to change. The pile will seem to be shrinking. Food scraps that were visible a week ago might have entirely vanished from sight. This is a good sign that your microbial community has begun to build. Decomposition has begun in earnest. Now is a good time to start using your nose. What you can smell is going to tell you a lot about what is going on in your pile. Bad odors are information. If you notice a stink, that most likely means you have too much nitrogen compared to carbon, so try mixing in woody material or torn cardboard and then leave the pile alone for a few days. Keep an eye on moisture overall. Don't be afraid to reach a hand in to "open" the pile a bit. If things are looking dry, add some water. If you notice food scraps that are lingering, like stubborn banana peels or that one whole onion you threw out, remove them from the pile, break them down into smaller pieces, and reincorporate them. See if that helps kick-start more active breakdown.

Around this time, you might also start to notice that the pile is warming up. One day, you may even see steam rising from the top. This is a good sign that the microbes in your pile are becoming abundant.

You'll also notice that you're attracting creatures. Worms, grubs, and beetles may start proliferating in large numbers. This is an excellent sign that the microbial population in your pile is stable and maturing. In my neighborhood, I tend to find fig grubs, thick as my thumb and almost translucent, along with slender salamanders, western blind snakes, and other creatures. Your local population of grubs and such may be different from mine, but once you begin to see them as positive signals of the health of your pile, they'll be no less delightful.

Now you are—officially—a composter.

# TO TURN OR NOT TO TURN

Just as there is no one right way to build a pile, there is no precise answer for when and how often you *need* to turn your pile. The "right" amount of turning is related to what kind of pile you're building, what conditions you're building it in, and when you'll need your finished compost. There are good reasons to turn and perfectly fine reasons to leave your pile alone. It's just a matter of understanding which method tracks best to whatever goal you have set for yourself.

### Why not to turn

Turning takes effort, physical capability, and time. You may not have those things. That's fine. Not every compost pile requires turning. Some methods even call for deliberately leaving a pile undisturbed for long stretches of time, which I'll address in later chapters. However, if you are not going to be physically turning your pile, you'll want to make sure you're mindful of finding other ways to maintain airflow throughout the heap. Poke holes with a shovel or insert thick branches, tubing, or anything else that will allow air to penetrate, continuously, deep into the heap. This will ensure your microbes are able to breathe, which will keep decomposition active.

### Why to turn

Turning your pile helps decomposition happen more quickly. It ensures continuous airflow for your microbes, even distribution of moisture, and more homogenous decomposition of your materials. Turning is a given in some compost setups, like three-bin systems, "hot" composts, and tumblers, but even single bins, open piles, and "cool" composts can still benefit from turning with some frequency.

## How much to turn

When determining how often to turn your compost pile, first consider your needs. Do you have the physical capability to turn? Do you want to run a "hot" compost? Do you need your finished compost on a specific timeline?

If you're not on a particular schedule, there's a lot less pressure to figure out the "right" turning frequency. Most unhurried backyard composters that I hang out with will turn their piles every few weeks, or even every few months, and leave it at that. If you are aiming to build a "hot" compost, you'll want to be more deliberate in your turning schedule. Those piles need to be turned every three or four days while their temperatures are high. They can be turned less once they start to cool. Hot piles are more efficient in their decomposition than cooler ones, so it follows that you might be looking to make compost quickly if you're running a hot compost.

Here are some other examples of when to turn—or not.

- **IF YOUR PILE IS BRAND NEW, DON'T TURN IT.** It needs time to get going. You can start turning it once it warms up or once you start noticing that things are actively breaking down and disappearing. The timeframe for this can vary, from a few days for a hot compost to a few weeks for one that stays cooler.

- **IF YOUR PILE BEGINS TO SMELL, GIVE IT A TURN.** This will break up any chunks of compacted material and reintroduce airflow.

- **IF IT'S WINTER AND COLD, TURN YOUR PILE LESS.** This allows the pile to retain any heat that it creates, which is better for decomposition. Some people even leave their piles unturned for the entire winter.
- **IF YOU HAVE A COMPOST TUMBLER, YOU WILL WANT TO TURN EVERY FEW DAYS.** This is because the enclosed environment of the tumbler can lack steady oxygen for your microbes, and turning helps maintain airflow.
- **IF YOU RUN A HOT PILE, TURN IT EVERY THREE OR FOUR DAYS WHILE THE TEMPERATURE IS HIGH.** Once it starts to cool and the materials have visibly broken down, you can stop turning.
- **IF YOU RUN A COOL PILE, TURNING ONCE A WEEK OR EVEN EVERY FEW WEEKS SHOULD BE FINE.** Just be mindful of airflow as you add new materials, making sure to incorporate enough bulky and high-carbon stuff as you build to create passive aeration throughout the heap.

If you do choose to turn your pile, your compost may benefit from a "curing" period once breakdown is complete. This means leaving it entirely alone for a matter of weeks or even months, during which the pile's microbial community will have time to rebuild and grow more diverse without being disrupted by turning. Just make sure you keep things continually moist while your compost is in this holding period.

Curing is optional, of course. If you need your compost right away, feel free to use it as soon as you determine it is ready.

# COMMON MISSTEPS

When people first start composting, there are a handful of mistakes they are almost guaranteed to make. These errors are understandable, often stemming from enthusiasm about the process or fundamental misunderstandings about how decomposition works, but they can lead a beginner to give up altogether. In the interest of ensuring that doesn't happen, I want to address at least a few of these most common errors head-on. (There will still be plenty of other mistakes for you to learn from.)

## Adding food items whole

While you can technically add whole food items to your compost, I don't recommend it. Instead, chop any food items into small bits before putting them in your pile. This ensures that they will break down quicker, smell less, and be less likely to attract animals. "Why would I even have whole food items to add to my compost?" you might ask. Just recall the last time a batch of bananas went bad on your counter, or the time you forgot about those three red peppers in your refrigerator. (This happens to me all the time.) Slice them up before you add them.

## Putting food directly on top of the pile

This is the one I see the most, and I personally think it's the most responsible for giving composting a bad reputation. When you throw food scraps directly on top of your pile and leave them exposed, you're more likely to end up with both animals and odors. Try your best not to do this. Every time you add food scraps to your pile, make sure you cover them thoroughly with high-carbon materials. Cardboard will suffice for this purpose, as will leaves, sawdust, wood chips, or a bunch of twigs and branches.

A COMPOST PILE BUILT INSIDE A SIMPLE WIRE ENCLOSURE ALLOWS FOR A GOOD AMOUNT OF AERATION.

### Keeping the compost in a small, dark enclosure

This is appealing for people who are squeamish about the idea of composting or find it "gross," but it often contributes to the very problems they are hoping to avoid. In my experience, compost enclosed in dark spaces is more likely to end up with imbalances, start smelling, and invite infestation from flies. This is simply because small, dark spaces make interacting with your pile more difficult to do, and thus you never learn how to manage it intuitively. These spaces also tend to lack good airflow.

PART TWO

# SPECIFICS

# FOOD SCRAP STRATEGIES

For many home composters, food scraps are a primary source of materials. However, I have seen more than one compost abandoned outright because its owner never managed to develop the habit of collecting their scraps. That's why I always recommend dedicating at least a few moments to thinking about both how you plan to store your food waste and how you expect to get it out to your pile.

Personally, I've always loved the open bowl. This is not an intuitively popular strategy, but there are a few reasons that it works so well for me. For one thing, I cook a lot and I mostly eat vegetables. My food scraps tend to be a harmless accumulation of onion skins and chard stems, zucchini bits and coffee grounds. Sometimes a bunch of flowers will end up in there. I also rarely have leftovers that include cooked food. I mostly tend to cook and eat in single-portion amounts. My compost pile is easy to access and I enjoy going out to see it, so the bowl gets emptied often. It is no chore to do so. And because the bowl is made of stainless steel and is easy to rinse out each time it's emptied, it stays perpetually clean—no gunk accumulates and there are no drips or congealed goo. Finally, the bowl fits into my cooking routine with ease and precision. It slides from the countertop to the butcher block to the kitchen shelf with no additional effort, meaning that I can whisk it over to where I'm preparing vegetables for dinner and then place it out of the way again when I'm done. I can leave it anywhere, and it fits nicely and looks fine sitting out in the open, in my opinion.

However, it's worth admitting that I also *like* looking at my food scraps. The bowl, as it fills slowly, becomes a pleasing snapshot of care. This is the food that I've cooked for myself and my friends. That bit of potato was from my garden, and I roasted it with chicken. Those three or four teabags are from when I had the girls over for dinner. The stems from those herbs went with eggs that I made for breakfast, and they made everything taste so

delicious. Any time that things in the bowl look particularly various (garlic skins, carrot fronds, bread crusts), I feel a little squirm of joy. I ate well this week. I fed people whom I love. Things can be simple and even good. The open bowl reminds me.

There's a lesson here, of course.

People often approach me asking for the "right product" to buy for their food scraps, when the solution is almost always to find the right *process*. The open bowl works for me because it suits my sensibilities, and my routine accommodates its best use. Your best receptacle will similarly fit your own preferences and habits, so start from there. How much food waste are you generating each week? How much space do you have? Are you willing to spend money? Do you mind looking at food scraps? These are all important. Once you have the answers to these questions in mind, it's easier to figure out how to go about things.

If you don't want to add extra steps to your routine, you could try replacing your regular garbage can with a double-bin system that lets you dispose of food scraps and traditional waste all in one place.

If you have a big family and don't want to spend a lot of money, a simple 5-gallon bucket might do the trick. I recommend using these in combination with some prepurchased sawdust or chicken coop shavings. Add handfuls each time you add food scraps to help manage odors and mitigate the potential for insect infestation. You could also freeze your scraps until they're ready to be added to the pile. Use reusable silicone bags that you can seal and store in your freezer.

If you drink mostly coffee at home and rarely prepare meals, you probably don't have to worry about having much food waste and basically anything will work for you. Just find any container. I'm including this option because, from what I can gather, this is true for a surprising number of you.

Truthfully, you don't have to spend a cent on a container if you don't want to. There are plenty of empty vessels drifting around the universe of any home that would suffice to carry future compost. Old canvas tote bags have worked for people, as have large jars and storage containers, repurposed take-out boxes, and paper bags from the grocery store. Anything that your food scraps fit in will work. Often, the biggest obstacle is just accepting that it's okay to improvise; finding a way to *enjoy* a new habit is the best way to make sure that it sticks. A friend of mine recently sent me a photo depicting seven or so different containers, from plastic take-out tubs to open soup bowls, all brimming with food scraps. "Would it kill me to have one normal-size compost as opposed to a million small vessels in my fridge?" she wrote.

I laughed, of course. So often you are doing the right thing with compost simply *because it is the thing that works for you*. There is no other standard you're being held to, which makes compost a refreshing rarity in life.

# CONSIDERING TOOLS

One of the great joys of compost is improvisation. That includes the materials that go into building your compost bin, what goes into the actual pile, how it gets there, and how the pile is managed over time. You can make things up, and anything goes. You may want to improvise from time to time, particularly as your intuitive understanding of the process grows. For this, you may want to have a basic set of tools around.

Tools can help you take quick advantage of some used wooden pallets or a pile of scrap wood. They can help you prep materials for your pile, and they can make the difference between a couple sweaty hours spent turning your pile and 10 short minutes. If you hope to become a long-term composter, that might mean a lot to you. Also, if you're like me, having good tools just makes you feel good about doing the work.

There's a suite of tools that I like to recommend to beginning composters, although I acknowledge that each item may not be equally useful to all. What you want to get will depend on what your specific approach to compost is going to be and what things you feel most comfortable using. It also matters what type of material you anticipate adding to your pile. For example, you wouldn't likely need a pair of hedge shears if you won't be clipping and collecting yard waste. Similarly, you wouldn't need a compost thermometer if your compost will live indoors and be too small to get to temperature.

## The right shovel

The right shovel varies based on what kind of compost you're making. If you are building an indoor or under-the-sink compost, you're likely working with a small pile that won't need more than a hand trowel. However, if you're dealing with a larger heap that is located outdoors, a good shovel is an absolute must. It will be your best friend, your right-hand guy, the thing that becomes almost an extension of your physical body.

Good compost shovels, in my experience, are flat shovels. They have a square blade that is cut straight across at the bottom. These types of shovels are excellent for scooping and turning, and the blunt end is also great for cutting through stringy yard waste. They are not always the best for digging actual holes in the ground, but they are excellent for scooping and turning compost. You will also want to look for a shovel with a handle length that works in proportion to your height to maximize your leverage. Trust me, this can make a huge difference.

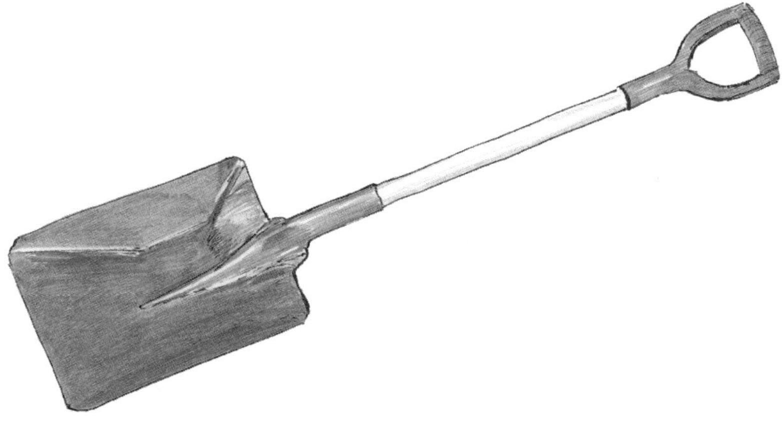

## A pH monitor

A pH monitor is not required in order to build good compost, but for the curious and methodical, it can be a useful tool for learning and monitoring. Mature compost usually has a pH between 6 and 8. When measuring pH, you should collect samples from throughout the pile to gauge the overall conditions. At-home monitors run about $20 and can be picked up at most garden or hardware stores.

## A compost thermometer

While you're definitely *able* to track the general temperature of your compost with a tool no more sophisticated than your own hand, I have grown fond of the humble compost thermometer over the years. Equipped with a long, stainless-steel probe and an easy-to-read monitor that categorizes different temperature ranges as "warm," "active," or "hot," this simple device allows the beginning composter to gauge the activity of their pile in just a few seconds. The general rule? The hotter the temperature, the more aerobic bacteria at work.

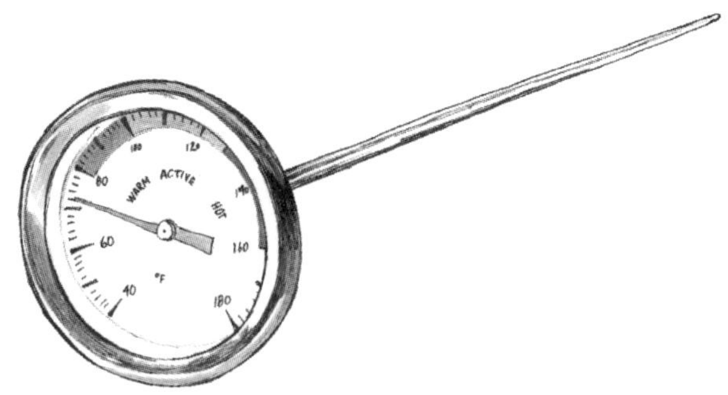

One of my all-time favorite activities is to plunge a compost thermometer into the heart of a pile, then watch the needle twitch and start to move, steadily rising toward higher and higher temperatures. (The highest I've ever clocked is 185°F, but temperatures that hot are not necessarily a good thing.)

## Hori hori

The hori hori is a traditional Japanese digging knife, featuring a blade of wonderful versatility. A little longer than 6 inches, it is thickly serrated on one side, straight on the other, marked like a measuring stick, and concave to accommodate digging.

I've lost parts of two fingers to my hori hori in separate incidents, but I retain my loyalty to it as an instrument. It can chop food, slice fibrous yard waste, and dig a modest hole. I've used it for prepping materials before adding them to my compost, turning the compost itself, planting vegetables, weeding, and even once for hacking a path through an overgrown garden.

If you had to pick just one tool from this entire list to buy, I'd tell you to make it the hori hori. Just be careful, because they are very sharp.

### Hedge shears

If you are composting in a yard, you want to be equipped to deal with yard waste. A pair of long-handled hedge shears will help you effectively snip back hedges and branches to add to your compost pile. I recommend trimming everything in your yard at once, piling it into a heap next to your compost, and then going to town on the heap with your shears. Just clip, clip, clip away. Once you've reduced everything to a fraction of its original size, it's generally ready to add. Don't overthink it, though. A mixture of sizes will be good for promoting aeration throughout the heap.

### Gloves

I never wear gloves when I compost, but most people prefer them. You want ones that fit snugly and, importantly, do not interfere with your ability to feel-by-touch as you work. Touch is a critical sensation for navigating your pile, communicating all kinds of information to you, like how dense or moist or dry things are. Thin but durable gloves, like ones made from rubber, are the most effective at shielding your hands from slop without completely blinding your senses as you work.

### Rubber gardening clogs

Composting can be dirty work. For some of us, it involves wading through swamps of partially decomposed food and mud, then tracking that through enormous piles of sawdust. This can be a huge nightmare for footwear, in addition to any cars or homes you might be subsequently invited into. Slip-on rubber gardening clogs are thus revelatory. Just grab a hose, rinse them off, and you're good to go again. The ones I wear were originally designed to be easily sterilized in medical environments. They're perfect for composting.

# DEALING WITH ODORS

As you learn how your pile works, how to maintain the right balance of materials, and what your pace of decomposition is going to be—you might end up with some odors. That's okay. For a beginning composter, it's even kind of a given. Consider it a rite of passage.

Luckily, bad smells are great teachers. They let you know when something is amiss in your pile and invite you to correct it. With time and practice, you will find that you experience fewer and fewer of them. You may, eventually, even learn how to prevent them altogether.

The number one reason that a compost begins to smell is that you have too much high-nitrogen material in your pile. When this happens, that excess of nitrogen ends up off-gassing as ammonia, which smells pretty bad. Another common reason is compaction. If your pile is too big and gets too heavy, it could compact under its own weight. When compaction occurs and airflow throughout the pile is eliminated, it causes your pile to go anaerobic, which means "without oxygen." Aerobic bacteria, which typically do the work of decomposition in your pile, are replaced by anaerobic bacteria. These decomposers break things down a bit more slowly and they also release nasty-smelling gas as they do. Your whole pile could go anaerobic, but the more common scenario is that anaerobic activity happens in patches. Both scenarios can be odorous. Luckily, though, both are also fixable.

If you notice an odor, a good and quick first action is to simply turn the pile. This can provide a helpful reset by mixing materials a little more evenly and reintroducing consistent airflow. If the smell persists, though, it might be time to take a step back and perform a more thorough intervention.

You'll want to do something I call "opening up the pile." That means dismantling the heap completely and then putting it back together while incorporating additional high-carbon materials throughout. Look for stuff that's sturdy, like wood chips and twigs, which can provide the balance you need and also create space for air to penetrate. Keep an eye on the pile's overall moisture levels as you do this. Add a little water if the final mixture appears to be on the dry side.

If you're dealing with a particularly bad degree of smell and turning your pile is not working quickly enough, you could also try adding a thick layer of sawdust or wood chips to the top of everything. This is called a "biocap." Keep adding material until the smell is eliminated, then leave your pile alone for a week or so.

If bad smells remain a problem for you, despite your best efforts to turn or add additional high-carbon material, you may want to be more strategic about how you're adding new stuff to your pile. Make sure you're not adding whole foods. Chop things down into parts. If they're already in parts, chop them down into smaller ones. This maximizes the surface area that microbes can access, which speeds along the process of decomposition. The quicker things break down, the less likely they are to smell. Some families I know make diligent use of food processors for this purpose, choosing to pulse their food scraps a few times before adding them to the pile. Make sure you cover any food scraps or other high-nitrogen materials with a fresh layer of high-carbon stuff. You can use sawdust, cardboard, twigs, or leaves. Whatever excess material you have on hand will suffice.

Over time, you'll learn how to read your pile with more fluency. You will be able to gauge the pace of its decay and add materials proportionally, and you'll likely notice that you encounter fewer and fewer odors. Pretty soon, you might even begin to realize that your compost smells *good*.

# DEALING WITH BUGS

Many people I talk to about compost say they are wary of building a pile because they're squeamish about bugs, or they've always wanted to try and can't quite bring themselves to because of bugs. In fact, I would cite "bug avoidance" as one of the primary reasons that most compost-curious people never get around to actually composting.

I don't blame anybody for their concerns. There are some solid reasons to be wary of bugs. Some of them harass you. Others will bite you. Many of them congregate where things are rotting, which is a bad sign if what you're looking for is a meal. However, I like to remind people, what you're doing is building a compost pile. Stuff is decomposing in there. You *should* see bugs in your pile, and I'd be worried if you didn't. Bugs are a good sign that your compost is working, that you have a thriving population of microorganisms, and that you're steadily achieving nutrient density.

However, acknowledging that bugs are a healthy part of the process does not necessarily mean that you're required to put up with an endless and overwhelming amount of them. The key to a good compost remains *balance*, and if you're seeing an extraordinary proliferation of one insect type, they may—similarly to a bad smell—be a signal that something in your heap is out of sync.

There are three types of bugs that seem to cause particular anxiety for people: flies, maggots, and cockroaches. None of them will harm your compost, but I acknowledge that they can frustrate and worry the composter. Let's talk through what each means—and doesn't mean—and what their presence might be telling you about your pile.

## Flies

There are two types of flies that tend to crop up in compost piles: fruit flies and houseflies. Both tend to show up where there is a lot of exposed food waste. The best way to combat that problem is to remake the conditions of your pile. Add more high-carbon material and turn your pile, ensuring that any food scraps are covered or buried. You might also consider adding materials that will make your overall compost slightly more acidic, which flies do not love. Think pine leaves or coffee grounds.

Overall, the best way to deal with flies, though, is to avoid attracting them in large numbers in the first place. Don't leave food scraps exposed on the top of your compost pile. Doing so will attract flies very, very quickly. If you add food waste, make sure to bury it within the pile or cover it with a fresh layer of high-carbon material. Keep an eye on your overall balance of carbon- and nitrogen-rich materials, and make sure you're not wetting the pile down too much. It should be moist, but not soggy and leaking water.

## Maggots and grubs

Maggots and grubs are different things. Maggots become flies and grubs become beetles. What they have in common is that they disgust people, although neither is bad for your compost. Maggots and grubs, by feeding on your compost, help break things down for you. Their numbers may surge if you add a ton of food waste at once, but they will naturally ebb once they've eaten everything. If you are seeing a ton of maggots in your pile all the time, though, it may be a sign that your overall ratio of carbon to nitrogen is consistently off. You should consider adding more high-carbon material, more regularly, to maintain balance.

### Cockroaches

It is inevitable that you may find cockroaches in your compost from time to time. We are so accustomed to their presence being a bad sign, some harbinger of doom, that we may react strongly. However, a few cockroaches are just another sign of decomposition-in-progress. They will break matter down in your pile, producing nutrient-rich poop in the process. In fact, their guts are such pathogen-busting powerhouses that people have begun experimenting with making whole composts out of cockroaches, similar to worm bins. You may not want that in your own backyard, of course. If you are suffering from an abundance of cockroaches, this—just like profusions of other insects—may be a sign that the balance in your pile is too nitrogen-heavy. You can even things out by mixing in high-carbon material. You might also want to clear away any nearby hiding places for them, like stacks of old wood or piles of leaf matter.

# DEALING WITH LARGER CRITTERS

A hard truth: There is simply no way to guarantee that an outdoor compost pile will never attract an animal. Not that I haven't seen many composters try their hand at it. They construct bins from sturdy materials with heavy lids, employing numerous locks and various design tricks, all in an attempt to dissuade the curious critter. Mostly, they only succeed in making their compost that much more difficult for themselves. Animals always seem to figure it out.

Part of composting is accepting this. You might get a skunk or a raccoon digging around on occasion, you might come out one day to find a rat has made a home in the upper layers of your heap. This is just part of sharing the world with other creatures, though, and I would encourage you to think of it as a relationship to negotiate, not a dire threat to eliminate.

What choices you make with your compost will depend on what animals are in your area, of course, in addition to your own comfort levels with their existence in proximity to your home. If you live in a city, rats are a known concern, and most people will

want to take reasonable measures to dissuade them from intrusion. Some folks are outright rat-phobic, and that has led me to give the rare advice that backyard composting might not be right for them. If you live in the woods, bears might be an issue. This is a potentially serious concern and should be treated as such. In my experience, raccoons and skunks tend to come and go more amicably.

In general, your best weapon is always knowledge. Learning to manage your pile efficiently helps deter interest, as does getting to know the habits and preferences of different animals in your area so that you can be careful not to inadvertently create their ideal home. Just know that no matter what you do, you may still get the occasional visitor.

### Take time to prepare your food scraps

You want to minimize the time that food scraps are sitting around in your pile, so anything you can do to expedite decomposition helps. Chop and shred things into the tiniest possible pieces before you add them. Stuff that's already small? Cut it smaller. On that note, and I wish I didn't have to say this so much, but *do not* throw whole food items into your compost if you are trying to avoid attracting animals.

### Cover your food scraps

Do not leave food scraps exposed, ever. Bury them in the pile. You might also employ a biocap, which, as I described earlier, is a thick layer of high-carbon material like sawdust, straw, or wood chips. You might want to do this in addition to having an actual, physical lid on your bin. It just helps reduce odors, which are the primary thing alerting animals to the presence of your pile. Do not leave food scraps sitting around waiting to be added, either.

### Restrict your materials

If animals, particularly bears, are a big problem in your area, there are some food types that you simply want to avoid composting altogether, like cooked food, meat, skins, oils, fats, and dairy.

### Get to know your locals

Familiarizing yourself with the preferences of the creatures common to your area can help you build more strategic deterrence systems. Most animals look for regular food sources in low-disturbance areas where they can also find places to shelter. Therefore, keeping the area around your compost clear of debris, piles of mulch, tall grasses, or any other potential hideouts can help dissuade them from setting up shop. Other animals dislike certain odors or food types, like orange peels and dog fur or urine.

## Put bad smells to use

The odor of food may attract animals, but there are other odors that can deter them. The Vermont Fish & Wildlife Department recommends leaving an ammonia-soaked rag in a bucket near your compost to discourage bears from investigating things. Refresh every two to three weeks.

## Build your bin on a barrier

The previous tactics may be sufficient to manage your critter issues. However, there is one higher-effort intervention that may be worth it if you live in a particularly creature-dense region. Build your compost on a solid base. That could be a poured concrete pad or a wooden deck that is raised an inch or so off the ground. This will ensure creatures cannot burrow in from below, which is one of the sneakier ways they will find into your pile.

# DEALING WITH COLD WEATHER

In compost work, winter is for resting. As the temperatures drop, you'll notice that everything begins to slow down. Decomposition takes longer, the pile may get cooler, and even the composter can feel a little sleepier, less inclined toward the heavy labor that can accompany pile management.

That's okay.

Renewal will come in the spring, as it always does. The days will get longer, the earth will warm, and decomposition—with its accompanying cycle of regeneration—will begin again. Your compost will be fine through this period of reduced productivity. (I want to say that you will be too.)

When winter comes, some composters I know will simply cover their bins and leave the compost alone for the season. This strategy has trade-offs, like needing to find another way to deal with your food waste for a few months, but it works perfectly fine. The compost will "wake back up" in the spring and the composter can continue as usual. Others will continue adding to the pile as they always have, understanding that things will just be slower for a while. This strategy also has trade-offs, primarily that composting while it's cold out can be, well, cold for the composter as well.

If you do want to keep composting through the winter, the best approach will vary based on how cold it gets where you live and how much effort you feel like putting into the endeavor. For example, those who live where temperatures drop below freezing will need to take more extreme measures to keep their piles active. For those like me, who live in seasonless areas like Los Angeles, composting year-round is much easier.

In general, though, composting in cold weather follows the same rules as any other form of composting. Each time you add high-nitrogen materials, like food scraps or coffee grounds, add high-carbon materials to balance them out.

Alternate layers as you add. Make sure to always keep food scraps covered. When the weather begins to warm again, the decomposition rate of your pile will kick back up, and you'll be back to the business of breakdown before you know it.

If you want to be more deliberate in managing your pile through colder temperatures, though, there are a few tactics at your disposal.

### Insulate your heap

If you live in a place where there is a risk of freezing, you could consider insulating your compost for the winter. Surround the outside of your bin with straw bales or big bags of leaves, depending on what you have on hand. You can also scoop everything out of your bin and line the interior with a thick padding of sawdust, leaves, wood chips, or some other carbon-heavy material. This will let your compost retain the heat of its own microbial activity and will help stop it from freezing over, so you can keep using it throughout the colder months.

### Keep things covered

If it's going to be raining or snowing a lot, that excess moisture can stifle the rate of breakdown. If you don't already cover your heap, lashing a tarp over it or building an awning can help keep the overall moisture of the pile stable, even through storms.

### Keep things right-size

Compost piles that are too small will quickly lose any heat they generate to the surrounding atmosphere. That means they're more inclined to freeze. Bigger piles are better able to retain the heat generated by their own microbial activity. This becomes a virtuous cycle, with the heat created by the active pile then *keeping* the pile active over time. Keep in mind that the ideal size for a pile to be able to retain its own heat is 3 cubic feet.

### Invest in preparing your inputs

The more you chop and shred stuff before you add it to your compost, the more you help out the microbes in your pile. You're giving them more surface area to penetrate, more quickly. When keeping things going while cold and slow, every extra millimeter of exposed material counts.

### Don't worry about turning

It's cold. Your pile is breaking down at a much slower rate. There might be snow. It could be pouring. You can give your pile a vigorous turn once spring comes. It'll be fine until then.

### Accept slowness

No matter what you do, your pile is probably going to slow down once it gets very cold. Just relax, accept, and let it take the time that it needs.

# THE QUESTION OF COMPOSTABLE PRODUCTS

These days, there are many products on the market that advertise themselves as compostable. These range from the single-use basics, like take-out containers and utensils, to more unexpected items like makeup remover wipes and candy wrappers. Some call themselves "backyard compostable," others merely "compostable." Some boast third-party certifications and others are dotted with a line of text warning you that "facilities for disposal may not exist." (Hmm.)

In the broadest possible terms, labels like "compostable" and "biodegradable" are meant to indicate that a product, once discarded, is capable of breaking down without leaving harmful residues in the environment. However, specialized facilities are often required in order to achieve this outcome.

This puts the home composter in a tricky position. The way these categories are utilized can be unclear and inconsistent at best and outright misleading at worst. Just because a product can break down in theory doesn't mean that it actually will be able to in your backyard pile.

Generally speaking, there are two ways to definitively determine if any given product is home-compostable. The first is to toss it into your pile and see if it breaks down. The other is to do some very thorough research on a case-by-case basis, investigating what ingredients an item is made from and their expected end-of-life treatment. You may also seek out third-party certifications, which offer more transparency into specific materials and manufacturing processes, in addition to outlining the conditions—home or commercial—under which a product is meant to decompose. Avoid anything that feels and looks like plastic. It's likely that those items will require special facilities to properly break down. To help understand this on more practical terms, an overview of some commonly purchased products can be helpful.

### Pet waste bags

How well compostable pet waste bags break down in a home compost system is a subject of great dispute. Given the volume of these bags that would need to be added to your home compost if you were composting for a pet, I personally would opt against them. It's just a lot of plastic-like product for your pile to try to digest.

### Flatware

Compostable flatware has become very popular at both parties and in delivery bags from restaurants. For the most part, these pieces are not able to be broken down outside of industrial composting facilities. I once threw a "biodegradable" fork into a compost pile in the community garden where I volunteer just to see what might happen to it. Two years later, it's still in there—a little bent, but very much a fork. The versions that are authentically compostable are made from wood or bamboo. If you want to compost these in your backyard, it's best to break them up into smaller chunks before adding.

### Grocery bags

Compostable grocery bags can be made with a truly stunning variety of materials, from vegetable starches and lactic acids to cottons and fungi. If you are certain to buy ones marked "compostable" versus "biodegradable," you should be able to break them down in an active backyard heap, particularly if you shred them into tiny pieces first. It's worth noting, however, that compostable grocery bags—just like plastic bags—can cause problematic pollution. They are easily misplaced, blown about, and end up as litter. My recommendation is to avoid them as much as you can, period. Switch to found totes and other bag-like items that you will be able to use repeatedly, even for collecting your food scraps.

### Face wipes

A few years ago, it suddenly seemed like every makeup wipe in the beauty aisle had started advertising itself as "home compostable." These wipes are generally made from easily decomposed materials like cotton or wood pulp, so the claim feels accurate—but your comfort with experimenting in your at-home compost with them might be contingent on what you're wiping off your face. I have used them and composted them myself, though, and I can report that they do decompose in an active backyard pile within weeks.

### Take-out salad bowls

It can be difficult to gauge the origin of a restaurant's packaging, but there are clues that might help you determine if a bowl is authentically compostable. If it feels coated or looks glossy, that's a sign there might be plastic lining. You're better off tossing it in the trash. Bowls that feel rough to the touch and uncoated are likely to be home compostable. You can also look for whether they are outright marked "compostable" or carry any third-party certifications. To add them to your compost, simply shred them into pieces and then add them to your pile as you would any other high-carbon material.

In general, I believe that compostable products work best when we compost them ourselves, whether that's in our own backyards or by bringing them to a community compost station. I am distrustful of stuffing things into bins and sending them somewhere else to be dealt with. I also think it's important to keep in mind, though, that a product being marked as "compostable" is not necessarily synonymous with "good for the environment." Compostable products only live up to their promise if we *actually make the effort to compost them*. They're still landfill if you throw them in the trash. They're still litter if you throw them on the side of the road. They also shouldn't be an excuse to maintain a habit of mindless consumption.

By now, many of us are accustomed to having takeaway coffee and food every day, carrying groceries home in plastic bags, and drinking water from plastic bottles that we can then just toss away. These conveniences are enabled by the existence of single-use plastics. But in order to imagine a future where compostable products truly fulfill their promise of a more sustainable mode of living, I believe we're going to have to think beyond "single-use" altogether.

Compostable products still require land and water to grow the ingredients that they are made from. They still require energy to produce and distribute. These are all environmental costs that will grow with their broader adoption, and those costs matter.

Therefore, we should consider their usage thoughtfully. We should aim to buy less, reuse more, and—of course—compost the rest.

# COMPOST PILE DESIGNS

---

There are probably as many ways to build a compost pile as there are people in this world. That means that, no matter how much space you have or what type of composter you want to be, there is a compost setup that will suit you.

## Single pile compost

The single pile compost is exactly what it sounds like. It is an approach to making compost that consists entirely of building and maintaining just one single heap.

Most home composters will never need more than this. It is also a great place to start if you've never composted before. That's because the single pile is so adaptable to different circumstances. You can build it all at once or over long stretches of time. You can turn it once in a while or never. You can put it out in the open, enclose it in a bin, or dig a hole and cover it up. Every imaginable variation is entirely acceptable.

Overall ease and flexibility are the huge benefits of adopting the single pile system. The potential drawback is speed. You are not likely to get finished compost very quickly with a single pile. For many home composters and casual gardeners, though, that's perfectly fine.

If you decide to go the single pile route, you can always purchase a commercially made compost bin. However, I personally recommend building a bin from materials you already have around you. Feel free to get creative. You could pile rocks into a square-shaped perimeter. You could use a combination of cinder blocks and weathered two-by-fours. You could cover the top with plywood and weigh it down with a brick. You could make a simple enclosure using chicken wire and a few yard stakes thrust into the ground. I have a friend in Montana who took over an abandoned tractor tire in a field for his compost. Another friend just uses an old cardboard box. Over time, as the box itself begins to

decompose, she folds it in on itself and puts it into a fresh box, where she starts everything up again. I've seen bins made from fallen logs, repurposed tree pots, old trash cans, and every other imaginable item. The range of found items that can be co-opted as compost containers is just truly enormous.

If you endeavor to build your own single pile compost enclosure, arm yourself with these guiding principles to build your best possible bin—no matter what materials you end up sourcing to do it:

- **YOU WANT A STRUCTURE THAT SUITS YOUR SPACE.** If you have a small patio, a full-size bin might be too much. Even though an "ideal" compost size is around 3 cubic feet, smaller is better if it means your compost bin doesn't encroach on your other habits in ways that will become frustrating.

- **YOU WANT A STRUCTURE THAT'S STURDY ENOUGH TO STAND ON ITS OWN.** You might consider ways to layer materials or reinforce siding. Keep in mind that compost, as it piles up, starts to get heavy. If the siding of your bin isn't strong enough, you'll get bulging and possibly bursting.
- **YOU WANT A STRUCTURE THAT PROMOTES AIRFLOW.** Gaps in the walls of your bin can help with that, although you may want to reinforce them with hardware mesh to discourage animal intrusion. You could also stack wood, lean wooden pallets together, or pile bricks loosely.

COMMERCIAL COMPOST BIN

COMPOST INSIDE A TRACTOR TIRE IN MONTANA

- **VISUAL ACCESS TO THE PILE IS BENEFICIAL.** Being able to see and interact easily with your compost will help you learn what the pile needs and become more familiar with the pace and processes of decomposition. This will help you maintain a better and healthier pile, which will help mitigate issues like odors and animals.
- **LIDS ARE OPTIONAL.** If you don't want to deal with one, you can just make sure to add a thick layer of high-carbon material to the top of your compost each time you're done adding new stuff in order to suppress potential odors.
- **A "FLOOR" TO THE STRUCTURE IS ALSO OPTIONAL.** Having a hard foundation can be useful if animals like rats or gophers are prominent in your area, though.
- **CONSIDER YOUR OWN USE OF THE BIN.** If you have the space for it, you may want to consider making your structure spacious enough to accommodate easy turning of the material. Wide and flat, with room for things to "heap up," can be very nice. If it's too tall and narrow, it can be hard to get a shovel in there.
- **THINK ABOUT HOW YOU WANT TO GET THE FINISHED COMPOST OUT, IF YOU INTEND TO USE IT.** You may want to think about leaving one side of the bin open or creating some kind of small door at the base.
- **ABOVE ALL, DON'T BE AFRAID TO GET CREATIVE.** You can always make adjustments later.

You can also, of course, opt for no physical structure whatsoever. For example, you can dig a hole in the ground and compost directly into the earth. Holes require a little extra labor on the front end but have the benefit of requiring no further effort and leaving no trace behind when you're finished. Holes full of compost can also, over time, significantly improve the soil in your yard.

To build compost in a hole, really all you need to do is dig the hole. Once you do, you can start adding stuff. Make sure to monitor moisture levels as you add material and that you're mixing enough bulky materials in to promote airflow throughout. Then, re-cover it with earth when you're finished. You can dig one big hole that you use continually, or you can dig several smaller holes that you rotate between. I have a friend who composts by digging a succession of small holes in her backyard, using them one at a time. She gradually fills each with kitchen and yard waste, utilizing an old aluminum trash can lid as a cover as she does. Whenever a hole is full, she simply covers it over with earth and moves onto the next. It's honestly quite genius. If you opt for this route, just be sure to keep your holes a couple feet from any plants you might be growing.

You might also opt for the open-air pile, a personal favorite of mine. This calls for heaping the various ingredients of your compost into a large, exposed mound. You can start with a layer of twigs or branches for a structured base that will promote aeration throughout the pile, then add a layer of food waste. Layer on more high-carbon material like leaves or ripped up cardboard, then pile on more food waste, and so on. As with any other type of compost, you'll benefit from making sure your

TRENCH COMPOST FILLED WITH FLOWERS AND CITRUS PIECES

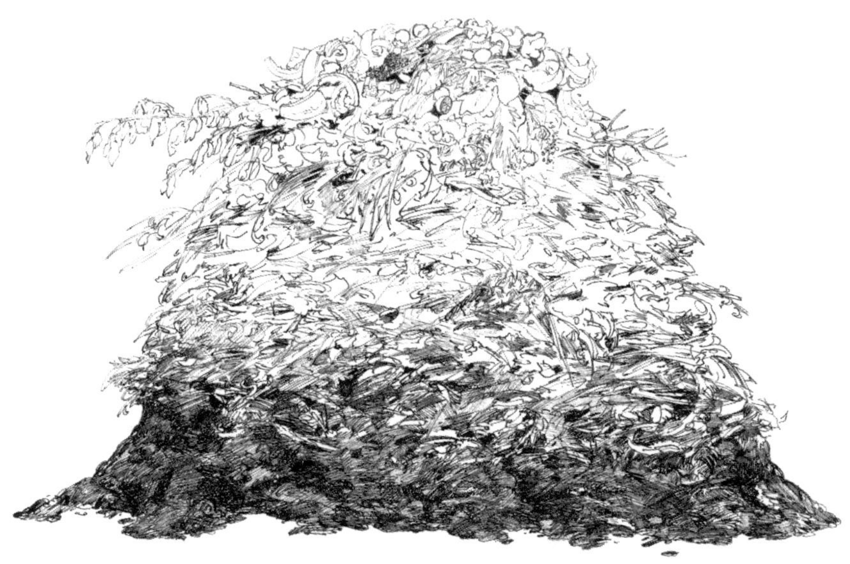

final layer is a thick coating of high-carbon material. This will smother odors and discourage flies.

Whatever structure—or non-structure—you opt for, removing the finished compost from a single pile system can be tricky. Just keep in mind that it doesn't have to be a precision effort. The compost at the bottom of your heap will be finished first. You can dig out a handful from the base from time to time to check whether the material seems fully decomposed. You can also turn the entire pile over every four or five weeks, so that the bottom becomes the top, then remove anything that looks like finished compost to you. Every pile will break down at its own rate, and you'll get a feel for how fast or slow your compost is decomposing as you interact with the pile over time.

## Three-bin compost

The three-bin structure is "the little engine that could" of compost. It's sturdy, efficient, and can make good compost quite quickly. Its basic components are, true to its name, three even-size bins that sit next to each other in a row. The front of each is made from removable slats or left completely open to make it easier to turn the compost between each bin. I've seen three-bin structures put to use in backyards, community gardens, and small farms. The community garden where I volunteer has two of them. With them, we are able to compost hundreds of pounds of waste per week.

The three-bin system is a wonderful tool when you're dealing with a lot of organic waste and also have a decent amount of time to dedicate to compost as a practice. It does require a significant amount of physical labor, so it is not ideal for those with limited physical capability.

TRADITIONAL THREE-BIN COMPOST SYSTEM

How it works is fairly straightforward but can take some practice to feel confident about.

Each bin is dedicated to a specific stage in the process of decomposition. The first bin is where the new pile is built, the second bin is where decomposition reaches its peak, and the third bin is where the compost completes breakdown. Your job as the composter is to turn the compost from bin to bin as decomposition progresses. This usually takes place across a period of about six weeks.

When getting started with a three-bin system, you'll first want to designate one of the bins as your "first." This will be where all your new materials are added. You should add materials in the same rhythm that you would when building any other compost pile, which means a solid base of bulky high-carbon material, like twigs or branches, then alternating layers of high-nitrogen and high-carbon items, keeping the mixture moist throughout. Some people fill the first bin all at once and others will fill it over time. Both are perfectly fine.

You'll know it's time to flip the contents of your first bin into your second bin when the first pile starts actively breaking down, becoming hot to the touch as it grows in size. Once you're done turning, though, you'll want to leave the pile in the second bin completely alone, so that decomposition can speed along without interruption. Do not add new materials to the second bin. During this time, the first bin, which is now empty, will be where you continue to add fresh stuff.

You will want to leave the compost in your second bin until the temperature peaks and starts to cool again. To track this, you can use a compost thermometer or your bare hand, although I would be careful with the latter. I've run three-bin systems where the compost gets up to 180°F in that second bin, which is hot enough to give you a decent burn. Once you feel that temperature start to come down, though, you'll want to flip the pile from the second bin into the third bin. At this point, you're likely ready to simultaneously flip the contents of the first bin into the second bin.

The third bin is where the compost will finish out the process of decomposition. You'll know it's ready when it's cool to the touch. It will also look and smell like good, rich earth.

When you really get things going in a three-bin system, every bin will be in-use at once. You'll be adding new materials to the first, scooping compost to use from the third, and checking temperatures on all of them, simultaneously. As you flip compost from one bin to the next, you can also use that time to check for overall moisture levels and aeration. If things look very dry, add water. If you notice a patch of stubborn and undigested food waste as you move material from the second to the third bin, simply toss it back into the first bin and send it through the system again.

You'll find the rhythm, and soon you'll be having fun. In compost, nothing is more immersive or enjoyably physical than the three-bin system. You can also manage a generous amount of waste easily, meaning you might even invite your friends and neighbors to contribute their food scraps.

Compost is, after all, best when undertaken as a community.

### HOW TO BUILD A THREE-BIN SYSTEM

The three-bin system can be built from anything. Wooden pallets can be screwed together and papered with chicken wire. You can stack bricks or cinder blocks. Most anything you find will suffice, that's the beauty of compost. However, if you are a proud home gardener or small farmer, you might want to build something that looks a little bit nicer.

If that's the case, there are many blueprints available for free on the internet. Just keep in mind that the cost of wood can result in some setups carrying a pretty hefty price tag. The three-bin design that I rely on utilizes a minimal amount of wood with just a few cuts, plus a few bits of hardware. The sides are reinforced with two-by-fours but leave enough open space to accommodate airflow. The backboard is extended in length to allow each bin to be a true 3 feet across. I don't include a lid unless large animals are a concern in the area where I'm building, to save on cost.

Remember that the three-bin system is generally made up of three bins that are each 3 cubic feet, which is considered the minimum ideal size for well-functioning decomposition. You could make the size of your bins larger, but I would recommend against anything bigger than around 5 cubic feet. The bigger your bins, the larger your pile and the heavier everything gets, making it more likely you'll deal with compaction and bad smells. Save yourself that trouble.

If you would like to improvise your own three-bin system, I wholeheartedly recommend rolling your sleeves up and getting into it. Here are some basic principles you should keep in mind, if you do.

◊ **EACH BIN SHOULD BE THE SAME SIZE,** somewhere between 3 and 5 cubic feet.

◊ **THE SIDES OF THE BIN SHOULD BOTH ACCOMMODATE AIRFLOW AND OFFER STRUCTURAL SUPPORT,** so there should be gaps—but also reinforcement. The gaps do not have to be huge; some ¼-inch slivers will do. More is fine, of course, but it does carry the risk of animal intrusion. If you have sides that are gapped enough to invite animal exploration, consider stapling hardware mesh or chicken wire over them.

◊ **THE FRONT OF THE BINS SHOULD "OPEN" TO ACCOMMODATE EASY TURNING.** This could be accomplished by building a front from slats or creating any other kind of removable door. Technically, you are not even required to have a front. You could leave the front portion of the bin open to the elements.

◊ **THE BOTTOM CAN BE OPEN TO THE EARTH,** but you may want to consider some kind of concrete pad if animals are an issue in your area.

◊ **LIDS ARE OPTIONAL.** You don't have to have them, but if you opt for a system without, consider keeping at least the first bin always covered with a thick mat of high-carbon material like leaves.

As long as your bins align with these basic principles, you should have a well-functioning system.

### WHERE TO PUT YOUR THREE-BIN SYSTEM

Deciding where to place your bins, given the size and general immobility of the structure, takes some special consideration. You'll want a place where the bins aren't cramped into a corner and there's at least some room nearby to temporarily heap materials as they're waiting to be added. You'll also want to be able to wield a shovel freely, so avoid areas with low-hanging branches or spiky bushes. If your yard offers barely any space, like mine, just do your best. Your best is enough.

## A word on efficiency

Some of the hardest battles I have ever fought in my life are with people who endeavor to overengineer a compost bin. They start simple, get excited, and then want to add a bunch of bells and whistles. "A bigger pile will be way cooler, wouldn't it?" they'll say. Or, "What if we made it look like a spaceship that spun continuously on a motor?" Or, "What if we . . ." etc. Contained within this enthusiasm is the well-intentioned desire to "improve" the compost through technological innovation, but the truth is that there is no real way to improve on the tuned-by-centuries process that is basic decomposition. The system must self-create to generate its own capacity. If you skip steps by introducing technologically driven efficiencies, you often end up paying the price later with an inferior finished product. Beyond that, if the project gets too complicated, it often ends up abandoned halfway through and then no compost gets made regardless.

Simplicity is always key. Stay the course.

## The lazy three-bin system

Building a three-bin system is not for everyone. That's why I sometimes defer to what I call the "lazy" three-bin system, which is a version that doesn't require any expensive equipment, hardware, or carpentry skills to set up. It's also highly mobile, in case you want to move it around, and it's easy to break down in the event that you decide you hate it.

To make your own, you just need a set of eight yard stakes, a roll of hardware mesh or chicken wire, and a pair of wire clippers. The stakes should be at least 4 feet tall. Place the stakes in rows of two, about 3 feet apart. Then unroll the hardware mesh and snake it through the stakes in an elongated "S." When you're done, snip the end of the hardware mesh or chicken wire free of the roll. You're done.

You can compost in the lazy three-bin system exactly as you would in a traditional system. Start with piling stuff into the first bin and then rotate it down the line. Because this system leaves your compost more exposed, you'll want to be mindful of covering the first bin and its fresher materials with a thick layer of carbon-heavy material to suppress any potential odors.

The lazy three-bin system works best for the composter who is not afraid to look at their own compost. While it is simple and cheap, it also can be leaky and exposed. If you build one and decide it's not for you, you can roll it up and take it away in a moment. However, you might find that you come to love your lazy little three-bin system. You might even find it beautiful.

THE LAZY THREE-BIN SYSTEM AT J'S HOUSE

I once built one of these systems for a private client at a residence alongside Griffith Park in Los Angeles. It was meant to be temporary, a way of experimenting with a location for her compost. I expected to use it for a few months before replacing it with a more traditional wood-and-metal one. However, the lazy three-bin system soon proved very popular. Every time I came to tend the system, staff and guests would float over to ask questions and admire the shaggy piles of plant detritus and finished compost. They liked being able to look at everything. So, the lazy three-bin system stayed.

# COMPOSTING INDOORS

In my experience, the obstacles to indoor composting are more spiritual than practical. Composting indoors costs little money, takes few materials, and can be done with relative ease, but would-be practitioners are dogged by some familiar concerns: odors, animals, and a general aversion to "grossness."

The difference for overcoming them indoors is primarily one of proximity. If you're worried about compost stinking, whether that compost is in your backyard or in your house could make a huge difference.

If this sounds like a familiar concern to you, it can be worth considering what things we already put up with from our garbage cans. They can smell. They definitely take up space. They get slimy and need to be cleaned out from time to time. Compost may come with some similar features, but the difference with compost is that all this potential nastiness is in service of things that are decidedly *good*. Preventing your food waste from being shipped off to a landfill is good. Turning your trash into fertilizer is good. Hopefully, this helps the reluctant indoor composter feel a little more excited about giving things a shot.

Of course, putting some time into thinking through your approach will help you get the best possible results—and result in the least amount of giving up. You'll want to consider how much and what *kind* of food waste you generate, in addition to how much space you have. Different setups accommodate different types of food waste and take up different amounts of space.

In general, the one universal rule of indoor composting is that you and your future compost will benefit mutually from an out-of-the-way place for storage, like a closet shelf or beneath a sink. This differs from my advice for outdoor composts out of plain practicality—limited space tends to mean that things get stashed unless they're getting used.

With all this in mind, let's walk through some of the options for indoor composting, so you can determine which suits you best.

## Bokashi

Bokashi compost is a method of fermenting organic waste in an airtight bucket, utilizing a special inoculant called bokashi bran. Benefits include containment—the bucket is small and the lid is sealed—and speed. The entire process is done in about six weeks. Bokashi can also handle meat, dairy, and bones. The cons are that the finished product is not *strictly* compost, although it can still be used as a soil amendment or added to an active

compost where it can finish breaking down. Depending on how much food waste you generate, you might consider purchasing multiple buckets if you go this route.

You can buy premade bokashi kits from a variety of sources. You can also build your own, although you'll have to purchase your own bokashi bran. You'll need two buckets and a tight-fitting lid to start. Drill holes in one bucket and set it aside. Take the second bucket and layer the bottom with cardboard or some other high-carbon, natural material. Paper or newspaper work. Fit the first bucket into the second, making sure to create a drainage area with something sturdy, like a few rocks or a wood block. Then add a few inches of food waste, followed by a sprinkling of bokashi bran. You'll want to sprinkle bran at frequent intervals to ensure its total penetration throughout your scraps. Mash everything down firmly in order to eliminate air pockets. (Remember: This is an anaerobic process!) Then put the lid on. Every few days, while your food scraps sit and the bokashi bran does its work, you'll want to lift the first bucket out of the second and empty the bokashi tea that has drained out. If you're very handy, you can affix a spigot to the bottom bucket to accommodate drainage. Within two to three weeks, your food scraps should be fully fermented.

Both your fermented food scraps and your bokashi tea can be used as fertilizer. The scraps can be dug directly into garden soil, although you'll want to add them at a distance of a few feet from your plants and ensure they are fully covered when you do. They can also be added to a regular compost heap, where they will complete breakdown and turn quickly into compost.

Bokashi tea should be used quickly, within a few hours of draining. It can serve as a nutritious drench for your garden or houseplants—just be sure to dilute it first, about 1 teaspoon per ½ gallon of water should suffice. You can also pour bokashi tea directly onto your compost to enhance your pile's microbial population.

### Worm bin

Commercially made worm bins that are appropriate for indoor use are abundantly available. These often look pretty in an apartment and can manage a reasonable amount of food waste—generally, worms can handle half their body weight in food scraps per day. The more worms you have, the more food waste you can process with them. The benefits of a worm bin are speed and that, candidly, worms are cool. They are sensitive to certain types of food, though, like citrus and tomatoes, and that may limit what you can compost with them. I'll go into this in more detail in the following chapter.

### The bag system

You can compost indoors by utilizing a pair of large, sealable plastic bags. Your initial build would be like any compost—a mix of high-carbon material with food waste—but instead of putting it outside, you'll shove it under your sink or anywhere else where it fits. You'll want to make sure oxygen remains in the bag, even after you've sealed it, in order for your aerobic microbes to survive. I also recommend finding some high-quality finished compost and mixing a handful in with your fresh materials. This will help kick-start decomposition by introducing a population of mature and diverse microbes. Keeping two bags allows you to have one bag for adding fresh material and one bag for letting things finish breaking down completely, without disruption. You can "turn" the compost in the bag where you are adding material by rolling it around every few days. Bags are nice because they're easy to carry when it comes time to dump your finished compost. They also are flexible and fit in more places.

### The bucket system

The bucket system is similar to the bag system but utilizes two 5-gallon buckets instead of plastic bags. This is the method I began using once I moved to an apartment. The bucket is lined with a grocery store paper bag to prevent odorous grime from accumulating on its sides. Food waste is carefully chopped down before being added. My high-carbon material is sawdust from a local woodworker. Every few weeks, I tilt the contents of one bucket into the other to "turn" things. This method is slow, though, and I don't recommend it for people with families or a large amount of food waste to manage. It works for me because I live alone and, thanks to my dog, have a very limited amount of leftovers.

### A note on countertop composters

Do your research. Countertop composters can sometimes overpromise ("Makes soil overnight!"), when in reality, all they're doing is grinding and dehydrating your food. That's not compost, although it can be a valuable service for making potentially smelly food scraps more manageable to deal with. Overall, though, I do not recommend them. They tend to be very expensive to only get a quarter of the job done.

## Be vigilant

Composting indoors is doable, but it undoubtedly benefits from having a practiced hand. You should be careful to chop food waste down into the smallest possible pieces before adding it to your setup, and keep an eye on overall moisture. The proximity required with indoor composting means there is less room for error when it comes to things getting wet and leaky. Make sure you're always adding high-carbon materials to maintain balance, as an outbreak of insects, inspired by an excess of food waste, could be decidedly unpleasant if it's happening directly inside your apartment.

You'll also want to be vigilant about oxygen. Because your setup is likely to involve some measure of containment—like a lid or a seal—it's just a little too easy to end up losing airflow. If you use a bag that you can seal, make sure you've "puffed" it with oxygen before closing it. Open buckets occasionally. Expose things in a deliberate and consistent way. Your microbes need to breathe.

As you do begin to produce your finished compost, you'll need to figure out what you want to do with it. Living in small spaces, without guaranteed access to a garden or yard, can mean that requires a little creativity. Houseplants can benefit from a handful or two of compost from time to time but do not need a constant influx. If you do have an outdoor area, you might contribute your compost for the general benefit of its soil. You could also donate to a nearby community garden or check in with any friends who garden. Either may be grateful for a steady supply.

## The benefits of lazy composting

I met a book editor once who asked me about how to compost if you have lots of space but just happen to be lazy. Her current compost setup, she explained, consisted of just throwing things into a large hole and ignoring them. She called it the "pit of doom."

"Honestly?" I said. "No notes."

Lazy composting is sometimes my favorite kind, because it can lead to some pretty great compost. Leaving your pile completely alone for long stretches of time can allow for the development of something beloved in the world of composting: fungal content.

Initial decomposition in a compost pile is largely driven by aerobic bacteria. They are the first microbes to show up, and they multiply rapidly in the early stages of breakdown. As time passes, though, the microbes in a pile will begin to diversify and other organisms will begin to flourish. These include fungal hyphae. They appear first as spores, then as filaments so slight that they can't be seen with the naked eye. Over time, they mature into visible structures, including actual mushrooms.

If you are turning your pile frequently, you're inadvertently breaking up the progression of those fungal hyphae. Their growth has to start over again. That's one of the reasons that frequently turned and "hot" piles tend to create more bacteria-dominant finished compost.

However, if you leave your pile entirely alone for long stretches of time, you could end up with compost that is very rich in fungal content. From this view, the "lazy" composter might be considered something of a master.

### THE SUPER-LAZY, NEVER-MESS-WITH-IT, HIGHLY-FUNGAL-FUN COMPOST

There are tactics beyond lethargy that help encourage the growth of fungi in your pile, of course. Cooler piles favor fungi, as do those made from a majority of high-carbon materials.

To make a Super-Lazy, Never-Mess-With-It, Highly-Fungal-Fun compost, simply source as much high-carbon material as you can. Dried leaves, twigs, sawdust, and cardboard will all work. Fungi enjoy a high-carbon meal and will thrive in a pile where your ratio favors that element. You'll also want to collect a minimal amount of high-nitrogen stuff, like coffee grounds or food scraps, which will still be needed to help get things going. Don't overthink it. Heap everything you've gathered together into a large pile, aiming for a height and width of 3 to 4 feet and ensuring that the high-carbon material fully covers any food scraps. When you're done, you can encircle it all with chicken wire, giving it a nice and permanent shape, or leave it unenclosed and shaggy. You can cover the top with a tarp, weighted on the sides with bricks. You can also simply leave it open to the elements.

Then just leave it alone. Your simple task, for the next year, is to make sure that things stay moist and that airflow stays consistent. You can do that by walking by once a week or so and just poking around. Wet things. Shift things, if necessary. In roughly a year, some portion of that pile will have turned into compost, and there is a very good chance that compost will be very high in fungal content. This method is effective, albeit extremely unscientific.

In my opinion, that's what makes it great.

There are a lot of reasons to want a fungal compost pile, even a modestly made one. One is that it takes time, and like a fine wine, the results can exude the pleasure of rarity. Another is more straightforward. Fungi play a critical role in overall healthy soil function, helping to cycle nutrients, store moisture, and provide pathogen resilience. Some plants even prefer a fungally rich compost. If you can, building and using a fungal compost can be a great gift to whatever patch of earth you have a duty to steward.

# HOW TO COMPOST WITH WORMS

"Vermicompost."

"Worm castings."

"Castings."

These words are all just fancy ways to avoid saying the truth: worm poop. Vermicompost is worm poop. You make it by raising some worms and then cultivating the poop they leave

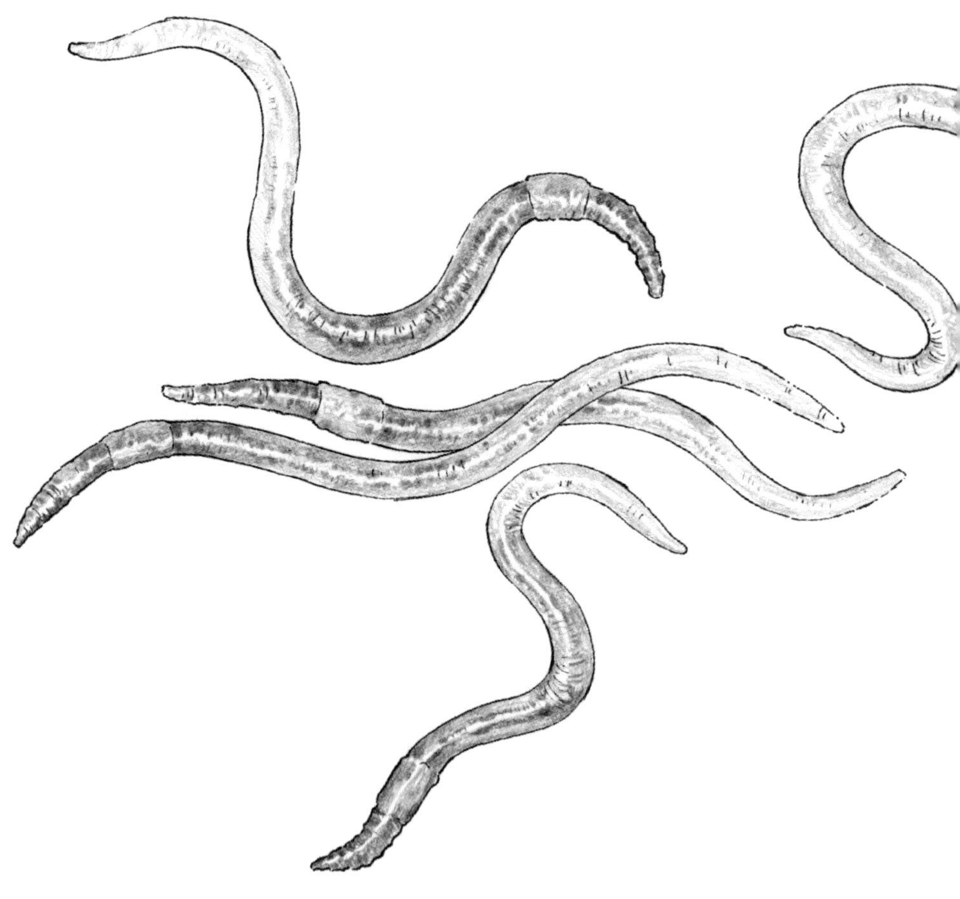

behind for your garden. Worm poop is considered to be one of the richest, most nutrient-dense amendments available for a farm or garden. Some lucky people will find an abundance of these creatures pooping freely in their backyard garden without any human interference, but many of us—facing increasingly degraded soils, urban pollution, and other obstacles—might need to give them a little assistance. That's where building your own vermicomposting system comes in handy. Luckily, it's easy. And as with most things compost, it can also be done for cheap.

### First, a note on worms

Not all worms are equal when it comes to compost. For vermicompost, red wigglers—the same kind you use when fishing—are the standard because of their vigorous and relatively undiscerning appetite. You can buy a lot of these guys for not very much money at almost any garden or fishing store.

In general, when you build any vermicomposting system, you're aiming to make an ideal home for worms. Worms love a little moisture and do not like sunlight. They need continuous access to oxygen, and they prefer shadier and cooler areas. They also benefit from some type of "grit" in their surroundings, which helps them digest things as they consume it alongside their food. Grit is a feature of most dirt, but if you want to be extra thorough, you can add a tablespoon of oyster flour or rock dust to your system. Understanding these basic principles of worm health will help you build a better habitat and take better care of them.

The vermicompost system that I learned to build is neither fancy nor particularly pretty, but it gets the job done and can be tucked away discreetly, if required.

### Building your vermicompost system

Materials:

- 2 plastic storage tubs
- A drill
- Dirt or compost
- Bulky high-carbon materials, such as cardboard, paper, eggshell cartons, etc.
- Water
- Food scraps
- Worms

To get started, pick one bin and flip it over. Drill a series of holes in the bottom. These holes are there to provide drainage for any food scrap leakage or for when you need to moisten the contents of your bin. Next, fill the newly drilled bin about one-third full with a mixture of dirt and bulky high-carbon materials. Whatever you use, make sure it's on the sturdier side. These materials are in the mix to retain moisture and create airflow for your worms. When you're finished, moisten everything down with a hose. Let any excess water run out of the bottom. You want things wet enough

DIY
VERMICOMPOST
BIN

that you can squeeze a palm full of the mixture and it will drip. You do not want it to be completely sopping wet.

Red wigglers will eat most things, including paper and coffee grounds. They will eat soft stuff first and then they will wait for anything else to break down until it's soft enough to eat. You can help them out a little by chopping up any food waste into small bits before putting it into the bin.

Add a few handfuls of food scraps, in proportion to the number of worms you have. You may only have a handful of worms to start, so you wouldn't need much more than ½ cup of food to feed them. As they proliferate, they'll be able to consume more and more food waste. You can increase poundage over time.

Once you are done building your worm bed, it's time to add your worms. Be gentle with them when you do. Nuzzle a hole in your materials, place them at the center, and cover them up again. Then, find a good piece of cardboard and soak it in water. Place this over the top of your newly built worm bed. It will help keep things dark and extra-moist.

You're almost done. Now, you will use your second storage bin as a catchment for anything that oozes from your first bin. You can DIY this by simply placing some large rocks or pieces of wood into the bottom and setting your first bin on top of them. This leaves some space between the two bins where liquid can pool, without drowning the worms. Just remember to lift the first bin out and check for liquid every few weeks. Cover your top bin and find a cool and shady place to store everything.

## Taking care of your worms

Similar to a standard compost, you'll come to learn the operating principles of your specific worm bin over time. You'll want to keep an eye on moisture levels. Don't let things dry out. You can add fresh bedding material, like torn up newspapers or more egg cartons, as you notice the old stuff disappearing into the bin.

Unlike a standard compost, your worm bins will require a little more care when it comes to what food waste you're adding and how often. You want to add food waste only in proportion to how quickly the worms can consume it. Add too much at once and they will leave stuff uneaten, which may then compact as it breaks down and potentially cut off their oxygen supply. Add food to the top layer of the bin—where it is at least partially visible. That way, you can easily track how much they are eating and when it is time to add more.

In terms of what foods to feed them, worms are more particular than your average pile. They are sensitive to things like citrus, in addition to acid-heavy foods like tomatoes. If you overwhelm them with either item, they could suffer. Avoid spicy stuff, dairy, meats, and precooked foods as well.

## Harvesting your worm castings

I have never been particularly scientific about collecting worm castings, something that shouldn't surprise you if you've made it this far in the book. I just dig around in the bin a bit, taking note of areas where the earth looks particularly dark and crumbly and there are no traces of food scraps. I'll scoop these out a few handfuls at a time and use them in my garden. If a worm or two comes along, that's fine by me. I am careful to harvest at least a few times a year, though, so that my worms are not living in a home that is made entirely of their own poop.

# CONSIDERING COMPOST TEA

Compost tea is made by steeping your compost in aerated water to extract nutrients and cultivate beneficial microbes. The finished product is extremely nutrient-dense and easy to spread in your garden. You can also tune your brew to encourage the growth of specific types of microbes, like bacteria or fungi, depending on the needs of your soil and plants. Think of it as a way to supercharge the capability of your compost.

Compost teas are cheap to make, and you can cook them up right at home. My own DIY setup cost about $40 between new and found materials: an old bucket, a repurposed barbecue grill, some mesh bags that are typically used to make nut milk, and a brand-new aquarium air pump and a few small air stones that I got from a local pet store. (You can buy commercially made brewers, of course, but they will be more expensive.)

**How to make compost tea**

You will need:

- Your compost
- Mesh bags
- Aquarium air pump and tubing
- Air stones
- Bucket
- Water

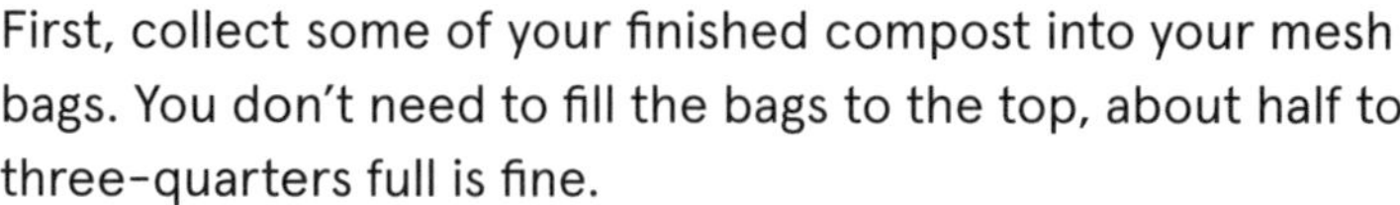

First, collect some of your finished compost into your mesh bags. You don't need to fill the bags to the top, about half to three-quarters full is fine.

Place your air pump and air stones at the base of your bucket. Fill the bucket about three-quarters full with unchlorinated

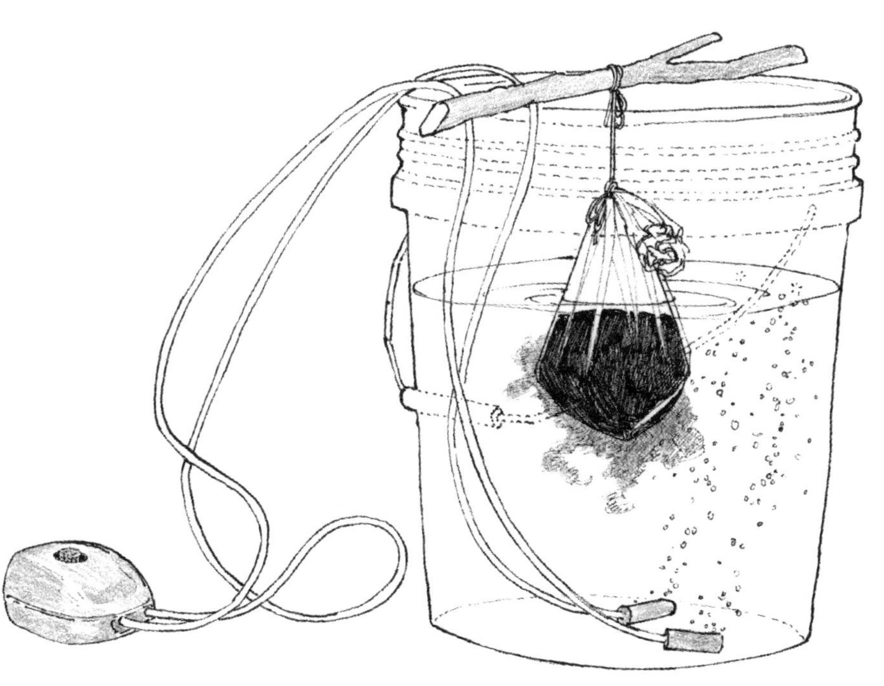

DIY COMPOST TEA SETUP

water. Turn the air pump on to begin aeration. Rest the bag of compost in the water, gently squeezing and massaging it, so that the bubbling water fully penetrates. I hang my compost bag from a repurposed barbecue grill, which I simply set on top of the bucket. I let the bag dangle without touching the bottom or sides. You might devise your own method for this, based on what you happen to have on hand. Let it all sit for 24 to 36 hours, then remove the bag of compost. What's left behind is your compost tea. It should be darkish brown and smell earthy and good.

### TWEAKING YOUR RECIPE

Based on what you're growing, you might want to build your brew to favor the growth of a specific type of microbe. Broadly speaking, do you want a compost tea with lots of aerobic bacteria? Or do you want one that has a lot of fungi?

If you're not sure what your plants would prefer, an easy shorthand is to think about their growth cycle. Annuals like vegetables, flowers, and grasses usually make better use of bacterially rich brews, while perennials like trees and shrubs prefer more fungal content.

Aerobic bacteria prefer simple sugars that are easy to break down. Think molasses and white sugar. If you want a bacterially-heavy brew, add the smallest amount—maybe ⅛ teaspoon—to your water as you begin aeration.

Encouraging fungi is a bit more complex. You'll want to get them started in the compost itself a few weeks before you make your brew. You can do this by taking several handfuls of your compost and separating them from your pile, then adding a fungal-friendly food. Think an ounce of fish poop or a small handful of worm castings. Powdered oatmeal works too. Keep an eye on things. Your compost will be ready to brew when you begin to see thread-like filaments, evidence that fungal growth has kick-started.

Here is a list of potential compost tea amendments and which type of microbe prefers each—although keep in mind you only need a very small amount of each to cultivate microbes in a five-gallon bucket:

**FUNGAL:**

- Fish hydrolysate
- Humic acids
- Kelp extract
- Seaweed extract
- Worm castings
- Powdered oatmeal
- Soybean meal
- Feather meal

**BACTERIAL:**

- Molasses (no sulfur!)
- White sugar
- Fruit juice
- Fish emulsion

It is important to note that there will always be many variables when it comes to the microbial population of your finished compost tea and very few guarantees that things will turn out one specific way versus another. How you make your own compost will be a factor, as well as what populations of microbes are already present. If you use worm castings, the diet of your worms will also be a factor.

The fun of compost tea, though, is the ease of experimentation. Feel free to mess around. If you have a microscope, or a friend with one, you can even look at your finished brew up close. Try to identify how many fungal spores you spot versus aerobic bacteria.

You'll know you've messed up if you end up producing a brew that smells, for lack of a better word, "rank." In those cases, something has gone wrong in the process and you're best off discarding the results.

### USING YOUR COMPOST TEA

You want to use your compost tea when it's fresh, ideally immediately after it finishes brewing or within a few hours, although it's best not to use it during the peak of the day, when evaporation will likely be quick. Timing your brew for early morning or evening dispersal is ideal.

You can use your undiluted compost tea as a drench for your soil or pour it directly onto the root zone of mature plants. I have never followed a particularly rigorous schedule to do this—about once a month has been fine.

To use your compost tea as a foliar spray, you should dilute it first. Half compost tea and half water, transferred to a spray bottle, and used as a slow mist on your mature plants will help introduce a diversity of good microbes to their leaves. This is like a quick-hit nutrient delivery for them.

Whatever you end up doing, though, just remember to pay attention to your plants. If you notice burns after applying a foliar spray, dilute your mixture more. If your plants are looking healthy and displaying increased vigor, you're on the right path.

PART THREE

# ENDINGS

# USING COMPOST IN THE GARDEN

If you're a home composter, you do not need any sophisticated methodology for determining when your compost is finished and ready to use in your garden (or, if you don't have a garden, ready to give away to your gardening friends). There are some basic questions you can ask of your pile to determine when it might be ready for use, however.

### What do you see?

The compost should be dark and crumbly, a lot like soil. Put your hand into your pile and sift it through your fingers. If there are still patches of slime or visible, "undigested" food scraps, your compost isn't ready yet. Let it keep sitting. If you have bulky twigs or wood chips remaining, though, that's fine. They can be sifted out before use, or simply applied with your compost—which is what I often do. What you really want to avoid is putting any undecomposed food directly onto your plants.

### What do you smell?

Compost that is ready to be used smells like fresh, good earth. It's not an odor you have to strive to perceive. Once my friend Vivian came over and stepped onto my back porch, just as—unbeknownst to her—I opened my pile up for turning. "What smells so amazing?!" she screamed. The odor of fresh earth had immediately and vibrantly permeated the surrounding air.

### What do you feel?

If the pile is still warm to the touch, that means that decomposition is still active, and your compost is not yet ready to use. Your best bet is to wait until it's cool to the touch.

OPEN-AIR COMPOST HEAP IN DAVID'S GARDEN

## How to prepare compost for use

Your finished compost may still contain certain visible elements, like twigs that are only partially broken down or the remains of bulky wood chips. This is quite common and can be fine, although some people wonder if they should sift these items from their compost before using it in their gardens. This one is tricky to answer definitively, but revisiting a basic overview of how decomposition, compost, and soil interact can provide some insight.

Microbes in your compost depend on a balance of carbon and nitrogen to process either element. When you maintain a compost, you provide the nitrogen your microbes need each time you add high-nitrogen materials, like food scraps or manure. However, if your finished compost retains visible pieces of high-carbon product, like wood chips or twigs, and you use that compost on your garden, your microbes may harvest what nitrogen they need to finish decomposition from your soil—possibly at the expense of your plants. If your overall soil health is very good, this may not matter. There's plenty of nitrogen to go around. If your soil conditions are degraded, though, competition for nitrogen becomes more meaningful. Your plants may suffer.

That's why some people will recommend that you sift your compost before using it in your garden. It's a safe way to ensure maximum nutrient availability in variable soil conditions. Sifting also has other benefits, like screening out stickers, rubber bands, and other small-scale garbage that can accidentally make its way into a heap. However, unsifted compost can also double as high-quality mulch and can be used effectively as such.

Like most things compost, there isn't a single correct answer. There's merely an assessment of conditions particular to your situation, plus a willingness to experiment and see how things work out.

Sift your compost when:

- You're generally unsure of soil health.
- Starting seeds.
- Fertilizing seedlings.

Don't sift your compost when:

- Preparing a garden bed in advance of planting, when you're reasonably sure that your overall soil health is good.
- Top-dressing healthy garden soil around mature plants.
- Top-dressing mature houseplants.
- Repairing bare soils.

Regardless of whether you choose to sift your finished compost or not, though, you should keep in mind that any compost needs to stay moist until you are ready to use it. Without consistent moisture, the microorganisms in your pile will die off or go dormant, leaving you with an inferior fertilizer.

## Considering compost and soil

Compost can do an extraordinary amount for your soil.

Its application will improve soil structure, increase nutrient availability, regulate pH, store carbon, and enhance water-holding capacity. Importantly, compost also brings with it a population of beneficial microorganisms. Once at work in your pile, they continue in your soil, feeding on organic matter and each other, reproducing and diversifying and continuing to make and cycle ever more nutrients as they do.

The beauty of this emerges as balance. Compost empowers a soil to both provide and to hold, releasing what is required—whether that's moisture or mineral content—at the time it is needed and storing whatever is not. Over time, that can make a soil quite independent, requiring less and less intervention from the composter to provide more and ever greater results.

One could compare that to chemical fertilizers, which offer a targeted abundance of a narrow range of nutrients to the soil, with the excess simply draining off. A soil's capacity for storage is never developed, and the soil remains dependent on the gardener to function. Not ideal—and, frankly, a lot of work.

If you'd like to use your compost to build your soil at home, it can help to first understand what type you've got. Luckily, soil is similar to compost in the sense that it can provide you with a lot of information through simple observation.

Take a look at the soil where you plan to use your compost. What color is it? Is it light or dark brown? Dig your fingers into it and take up a handful. This action also provides information. Does the soil yield easily to touch? Or is it hard? Once you have some in your hand, feel it in your palm. Take note of its texture. Is it gritty and sandy? Or is it softer and loamy? Do you see any insects in the hole left behind from where you dug out some earth?

Depending on what you want to use your soil for, these are all great data points to have—and you got all of them just from looking and touching.

Darker soils that are crumbly tend to be more fertile. You'll find that these types of soil tend to exhibit signs of life, like insects and worms, even before you start to plant in them.

Soils that are light-colored and gritty are called "sandy." They drain water quickly and tend to be lower in nutrients. Heavily compacted soils can be a sign of generally degraded conditions, and plants might struggle to take root and thrive. This makes sense, of course. If your hand has trouble breaking up the dirt, so would a plant's roots.

Knowing what kind of soil you are starting with will help you make a plan for how to best use the compost you are making. A few suggestions follow for how you might approach tending your soil, but keep in mind that they are not comprehensive.

Consider them a starting place and an invitation to be curious and keep exploring.

If you are lucky enough to be starting with a healthy soil, growing plants like vegetables and herbs is going to be much easier for you. Your main task will not be to improve your soil, but to help maintain its vigor. You can do that in several ways. Dig furrows alongside your garden and fill them with compost, wetting them when you're done. Add compost to holes before you transfer in mature plants or seedlings. After a final harvest, layer an inch of compost over your garden, cover it with mulch, and wet it

down. This will help restore microorganisms to the soil that were depleted from the growing season. You'll want to keep your compost/mulch dressing moist until you're ready to plant again.

If you have sandy or heavily compacted soils, it can take more time and effort to build them up. You can top-dress with a mixture of compost and mulch, like you would with more fertile soils, but this can take several years before you start to see an impact. More strenuous interventions can help speed things up a bit. For example, you can commit to a one-time tilling event, which calls for incorporating a generous amount of compost into the top 4 to 6 inches of your soil. If you have sandy soils, hand tools should suffice for this. More compacted soils will require hardier instruments, like a full-size shovel or a broadfork. Once you've dug in the compost, cover it with mulch, and water it all in heavily. The moisture will ensure that your microbes survive.

All that said, what makes for a "good" soil is a dynamic concept. Different types of soil are "good" for different types of plants and needs. You may not want to grow vegetables. Perhaps you like a garden filled with native plants or a mix of ornamentals and trees. If that's the case, achieving maximum fertility for your soil may not actually be ideal. Some native species in California, where I live, prefer a sandier soil and a lower-nutrient environment. If I wanted to grow them in my yard, I would keep my compost very far away from them.

## Using compost to start seeds

One summer, an old friend of mine moved to Los Angeles from the East Coast. I brought her five buckets of finished compost as a welcome gift. I thought she could use them as a base to start a vegetable garden, but she had a different inclination. "How do I use them to start seeds?" she wondered.

I will acknowledge that the conventional wisdom on using compost to start seeds is often "Don't." People will mention that the risks are too high, the trade-offs insufficient. They will often bring up the possibility of pathogens, fungal infections, or a nutrient overload that will harm fragile seedlings. I think that all these issues can be avoided if you know where your compost came from—in this case, yourself—and if you manage your seedlings with a minimum amount of care. In general, I like using my own compost to start seedlings just to keep things super local and DIY.

In general, you should use compost that is mature and sifted to start seeds, so that it's a bit lighter and less bulky. If your compost still has any visible material like food chunks breaking down in it, let it keep cooking before using it to start seeds. It's ready when it looks and feels like crumbly soil.

If you are starting seeds for growing food, it's also wise to consider the properties of your own pile to make an informed choice about what type of seeds are appropriate to use it with. Compost that comes from a "hot" pile, one that has reached temperatures of around 140°F, represents reduced pathogen risk and can be used more confidently. If you composted any diseased plants

in your pile (not recommended, generally, but it happens), you wouldn't want to then use it to start seeds. You risk passing along the infection. If your compost was on the cooler side throughout breakdown, you also might err on the side of using it only with seeds for inedible plants.

In general, seedlings require consistent moisture and access to oxygen to germinate. Compost tends to be denser and heavier than traditional seed-starting mixes, but you can lighten your batch by incorporating roughly one part coconut coir to two parts compost. The mix will be right when it feels light and fluffy to the touch.

You can repurpose containers around the house to start your seeds. Egg cartons are great, as are old newspapers, which can be folded into a cup shape using a water glass. Just make sure you poke holes in the bottom to accommodate drainage.

Add your compost to your pots with a light hand, making sure it doesn't tamp down and eliminate airflow for your seedlings. Check the overall moisture, as well. You want your mixture to be damp, but not leaking rivulets of water. You can pre-wet your mix accordingly, or water your seeds in after you've planted them.

Get your seedlings and tap them into the compost mixture. You want to push them gently down about ¼ inch. When you're finished, set them aside and wait. You should start to see growth soon.

When it comes to seedlings and compost, my opinion is that you should feel free to experiment and learn from what you see. If your seedlings sprout vigorously from your compost mixture, that's great. If they shrivel and die off young, you probably need to vary your mix. Try mixing in a little more coir or even regular soil. You'll also want to double-check drainage. If your compost mixture is waterlogged, you may end up with poor growth or dead sprouts. Ultimately, see what does and doesn't work for your own seeds and iterate accordingly. Learn as you go. Look and touch and smell.

I do like to start my seeds with compost, and I do it all the time—although not exactly on purpose. My vegetable scraps, like those from squash and tomatoes, tend to just spontaneously sprout from my backyard pile. I had a patch of my old garden dedicated to replanting the random stuff that was sprouting from my pile. In general, I do not particularly pamper my plants—I've always been a little devil-may-care. I do, however, like to provide them with a solid foundation, which to me means good soil. And, yes, that includes compost.

## Other ways to use finished compost

There are a few other ways you can put your finished compost to use. I offer some examples here, although your imagination may furnish you with other, even greater options.

### Perk up a houseplant

Mix a handful of compost into your regularly purchased soil when potting or repotting a houseplant. I've seen it recommended to use one part compost to three parts soil, but I'm generally imprecise and highly unscientific, so I just say "a handful." Make it two if the pot is particularly large.

## Share it with friends

This one might be my personal favorite: Give your compost away. Not all those who compost also garden. Maybe you live in an apartment without a yard or have a picky landlord who doesn't want you to touch their garden. Hey, it happens. I bet that your local elementary school, community garden, senior home, or whoever else would love a donation. Your neighbor might like some, your friends might like some—just ask around and see.

# A WORD ON COMPOST, OAK LEAVES, AND ECOLOGY

California is home to nine species of oak: valley, blue, coast live, Engelmann, canyon live, interior live, black, island, and Oregon white. Together, these stooped giants drop a thick carpet of sturdy, carbon-rich leaf matter beneath their canopies, providing the soil with a slow-decaying feast of critical nutrients.

Oak leaves can also make for excellent compost. They're nutrient-dense, carbon-heavy powerhouses that many people will rake up and add to their backyard piles, something that feels akin to "cleaning up."

However, the orderly backyard is often created at the expense of ecological balance—when you take oak leaves for your pile, you are taking them away from the oak tree. I often advise people to source inputs for their compost from their surroundings, but I also take pains to remind them that they're sharing those resources with their plant and animal neighbors. Be gentle with how you gather. For example, the oak tree really counts on its own leaf litter for a number of reasons, like nutrients, enrichment of the soil surrounding its roots, and the suppression of unfavorable plants ("weeds") that would disrupt its feeding and reproduction. Oaks are very sensitive to what plants they share space with. They thrive best alongside coastal sage scrub and chaparral, but these types of plant communities are vanishing and you're now more likely to find oaks surrounded by European grasses, especially in residential neighborhoods. As a result, oaks are dying across the state. Trees weakened by poor landscaping practices are less likely to flower and less likely to produce acorns. If acorns are produced, they're less likely to germinate and even less likely to successfully grow.

To understand the principles at play, it helps to understand a little more about how an oak "eats." Oaks are ectomycorrhizal, which means they utilize ectomycorrhizae (EcM) to help them find and intake nutrients. EcM form a sheath around the root hairs of the oak, surrounding the cortical root cells, and then search for food. They reach up into the oak's layer of leaf litter, where they source small amounts of nutrients made available by the slowly decaying leaves and transfer them back to the oak. This is a clever and symbiotic survival strategy, but it also makes the oak highly vulnerable to soil disturbance around its roots, such as from the removal of its fallen leaves and the subsequent encroachment of invasively behaving grasses that do not provide nutrients.

If you have an oak in your yard, you may feel suddenly moved to rush out the store and purchase a commercially available mycorrhizal inoculant to help it out. I would caution you against this. Most inoculants are made of endomycorrhizae, which is a separate type of differently functioning mycorrhizae. You might also worry that you need to fertilize. Don't do this either. Introducing external, synthetic fertilizers will cause the oak to disconnect its slow-built mycorrhizal relationships as it takes in a sudden excess of quickly available nutrition. You do not want to destroy that carefully built network. All you want to do is get out of its way and let it do its thing. So, as much as I love to compost everything in sight, I also always recommend that you leave at least some of your oak leaves alone.

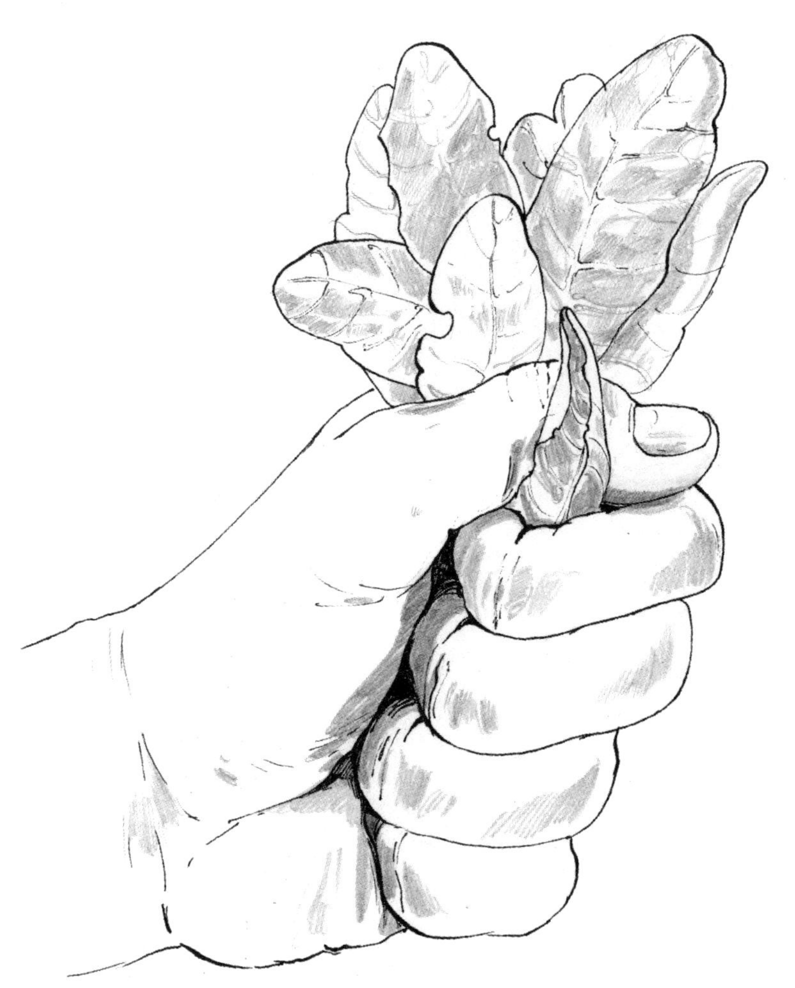

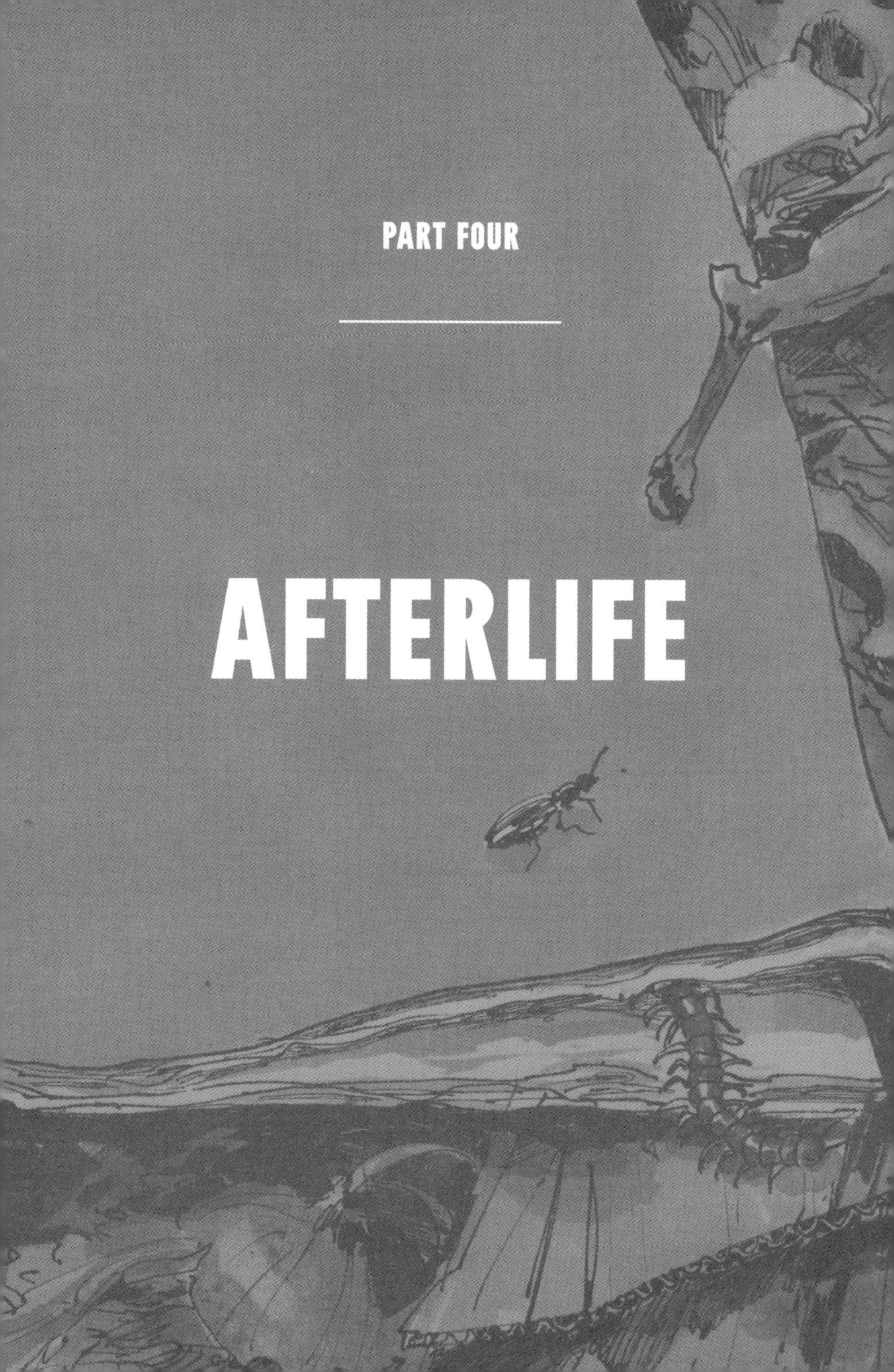

PART FOUR

# AFTERLIFE

# COMPOST AFTER READING

I hope you've learned what you need to know about compost. Maybe by now you've even made compost. Or if you've already been composting, you've made better compost, or convinced someone else to make compost, or shared your compost with your neighbors.

The only thing left to do now (besides making more compost!) is to compost this book.

This book has been printed on paper certified by the Forest Stewardship Council (FSC), and its pages will make for an excellent source of carbon in any compost pile. To compost this book, I recommend first destroying it. Separate the pages from the binding. Discard the binding, just to be safe. Now shred the pages into pieces. You might go page by page. You might grab handfuls and rip. You might use scissors. You might even—and this is for the both courageous and safety-minded—burn the pieces and turn them into ash.

Next, you can begin your compost. Moisten the paper and mix it with food scraps. If you've burned your book into ash, pinch it with your fingers and sprinkle it throughout your pile. Or add it all at once—who cares?

The knowledge from this book lives in your head now. No, not knowledge. Let's call it an invitation. I hope this book invites you to try and see what happens. Trust yourself to make mistakes and to learn from them.

Go on now, make compost.

# ACKNOWLEDGMENTS

There are an unbelievable number of people who facilitated the creation of this book in every imaginable way.

Thank you to my agent, Bridget Matzie, and to Felix Salmon for introducing me to her. Thank you to Makenna Goodman, my editor and now friend.

Thank you to the writers and fellows of Why Is This Interesting?, who have buoyed me every inch of the way with their unconditional support and enthusiasm.

Thank you to my best friend, Caitlin Kroeger, who inspired me to change my life and do something worthwhile.

Thank you to my family, specifically to Mom, who was my first composter, and Dad, who loves nothing more than cleanliness and yet still unequivocally supports my right to be constantly covered in dirt.

Thank you to Kevin Spring for being my biggest cheerleader.

Thank you to Martine Syms for being an unbelievable friend and creative mentor. You have inspired me beyond measure. Thank you to Rachel Knowles for always picking up the phone to call me when I sent what you knew was an anxious text.

Thank you to Lawrence Chit for volunteering to design the entire first version of this book to ever exist, after I couldn't—for the life of me—figure out InDesign. Thank you to Andie Marsh for your generous, beautiful mind and your willingness to always dialogue with me about soil and science.

Thank you to Anthony Tran for reading and editing early drafts of this book and for being willing to give me the tough, good feedback that I needed to hear. Thank you to David Horvitz for deciding I was an artist and then treating me like one.

Thank you to all the people who spent time talking to me and teaching me about compost: Noemi, Aaron, Sam, Lynn, Maggie, Nina, Dujon, Christian.

Thank you to all the people who spent time talking with me about writing and making work: Cyrus, Hurley, Amanda, Linsday, Molly, Andrea, and Sam again.

Thank you to The Rot Squad, who have composted alongside me every week for years. You have done more for me than imaginable.

# RESOURCES

Each of these resources influenced my work with decomposition deeply, but not because all of them are directly about compost or soil. Some taught me other, less direct but not less important, lessons.

*Soil is Sexy* (Substack), Andie Marsh – soilissexy.substack.com

*SoilWise* (Substack), Lynn Fang – soilwise.substack.com

*For the Love of Soil: Strategies to Regenerate Our Food Production Systems* by Nicole Masters (Integrity Soils, 2019)

*Dirt: The Ecstatic Skin of the Earth* by William Byrant Logan (Riverhead, 1995)

*The Rodale Book of Composting: Easy Methods for Every Gardener*, edited by Grace Gershuny and Deborah L. Martin (Rodale, 1992)

*A Darker Wilderness: Black Nature Writing from Soil to Stars*, edited by Erin Sharkey (Milkweed Editions, 2023)

*Thinking in Systems: A Primer* by Donella Meadows (Chelsea Green Publishing, 2008)

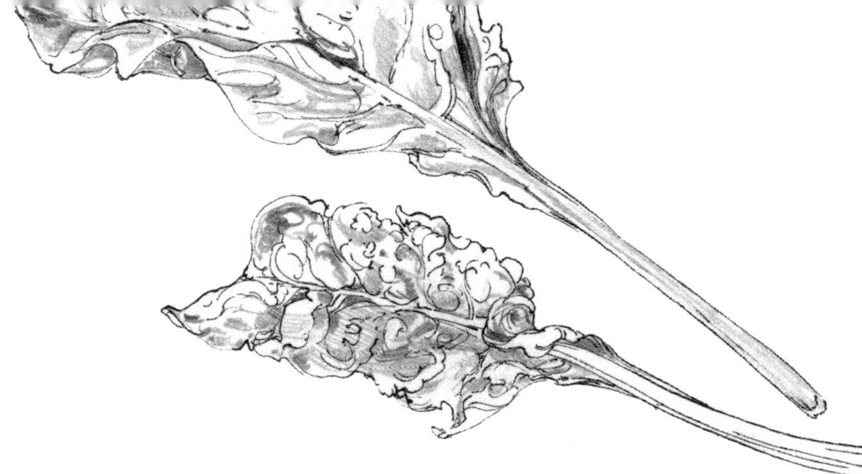

*Hospicing Modernity: Facing Humanity's Wrongs and the Implications for Social Activism* by Vanessa Machado de Oliveira (North Atlantic Books, 2021)

*Staying with the Trouble: Making Kin in the Chthulucene* by Donna J. Haraway (Duke University Press, 2016)

*Secrets of Fertile Soils: Humus as the Guardian of the Fundamentals of Natural Life* by Erhard Hennig (Acres USA, 2024)

*Teaming with Microbes: The Organic Gardener's Guide to the Soil Food Web* by Jeff Lowenfels and Wayne Lewis (Timber Press, 2010)

*New Weathers: Poetics from the Naropa Archive*, edited by Anne Waldman and Emma Gomis (Nightboat Books, 2022)

*A Billion Black Anthropocenes or None* by Kathryn Yusoff (University of Minnesota Press, 2018)

# INDEX

CAROLINE TOMPKINS

**CASSANDRA MARKETOS** is a Los Angeles–based community composter, writer, and artist. She works in her neighborhood to divert food waste from landfills, build and maintain composts with neighbors, and educate students on the principles of decomposition.